"This book is packed with delightfully fun projects for a rainy day or romping in the sun. The ratio of building and playing is just right."

—Chris Anderson

Praise for

Geek Dad:

"Includes scores of illustrated projects that a parent and child can do together, using many materials that barely existed a few years ago . . . Stand back and watch the fun ensue."

—*The New York Times*

"These are truly fun, inspired, and even educational projects you can do with your kids." —*Wired*

"There are projects that require computer skills or some knowledge of electronics; mostly, though, they require inquisitiveness and imagination . . . The bottom line—just as it was with our fathers and grandfathers—is doing something with your kids that is fun and interesting for all parties." —*Chicago Tribune*

"Projects for science enthusiasts of all ages."

—*NPR Science Friday*

"Call it *Revenge of the Nerds V*. . . . A book that embraces geek culture." —*San Francisco Chronicle*

"A how-to guide that includes nighttime kite flying, electronic origami, and cyborg jack-o'-lanterns." —*USA Today* gift guide

"A great book . . . This is fun you can have with your kids, with your coworkers, with your spouse, and your friends."

—Ars Technica.com

"[*Geek Dad* features] dozens of geek-friendly activities and rainy-day projects for parents (not just dads, we should note) and kids to enjoy, from creating your own comic strips to building a working lamp out of CDs and LEGO bricks. Pick it up in time for Father's Day, and be sure to let Dad know he has to share."
—Babble.com

"Read this crafty book for ideas to share your love of science, technology, gadgetry, and MacGyver. . . . Soon, together, you can rule the galaxy as father and son. Mwahaha."
—Boston.com

"The book provides an easy gateway to spending purposeful time with your kids and sharing experiences that they'll never forget, all in the spirit of tech-savvy DIY. Now that's an idea to geek out about."
—GearPatrol.com

"I think many geek dads (and okay, geek moms, too) will appreciate the hands-on approach that fills the void that kids just don't get today in school, especially in the American culture of 'teach to the test.'"
—GeekCowboy.com

"Learn to create all kinds of great geeky stuff and fun with your offspring, including the ultimate in summer fun—a super Slip 'N Slide."
—GeekyReader.com

"What a fun, educational but not in a boring way, book!"
—MegansMusings.com

"Full of all kinds of cool projects you can do with your kids, the book is a must-have for any dad (hint: perfect Father's Day gift!) who loves to get geeky with the little ones."
—MentalFloss.com

"Craft projects to be shared by kids and techie dads are collected by the author of the GeekDad blog on Wired.com. The best ideas involve the making of a comic strip out of LEGOs and flying a light-rigged kite at night."
—*Milwaukee Journal Sentinel*

"A compendium of fun and geeky projects for kids to do with their parents." —Makezine.com

"An easy to read, adroitly written craft book." —Neatorama.com

"I would encourage any adult with latent geeklike tendencies to acquire this tome to realize their full potential." —SmallTownLiving.com

"A perfect manual for sharing your geekiness with the next generation." —ThatsBadAss.com

"Denmead's subcategories for each project—concept, cost, difficulty, duration, and reusability—help readers easily navigate the book to find items they want to make that match their budgets, skill levels, etc." —UrbanBaby.com

GOTHAM
BOOKS

Guide to Weekend Fun

Cool Hacks, Cutting-Edge Games, and More AWESOME PROJECTS for the Whole Family

KEN DENMEAD

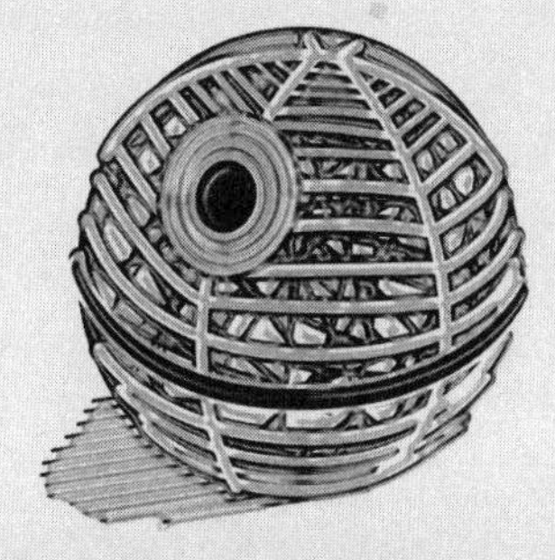

GOTHAM BOOKS

Published by Penguin Group (USA) Inc.
375 Hudson Street, New York, New York 10014, U.S.A.
Penguin Group (Canada), 90 Eglinton Avenue East, Suite 700, Toronto, Ontario M4P 2Y3, Canada (a division of Pearson Penguin Canada Inc.); Penguin Books Ltd, 80 Strand, London WC2R 0RL, England; Penguin Ireland, 25 St Stephen's Green, Dublin 2, Ireland (a division of Penguin Books Ltd); Penguin Group (Australia), 250 Camberwell Road, Camberwell, Victoria 3124, Australia (a division of Pearson Australia Group Pty Ltd); Penguin Books India Pvt Ltd, 11 Community Centre, Panchsheel Park, New Delhi–110 017, India; Penguin Group (NZ), 67 Apollo Drive, Rosedale, Auckland 0632, New Zealand (a division of Pearson New Zealand Ltd); Penguin Books (South Africa) (Pty) Ltd, 24 Sturdee Avenue, Rosebank, Johannesburg 2196, South Africa

Penguin Books Ltd, Registered Offices: 80 Strand, London WC2R 0RL, England

Published by Gotham Books, a member of Penguin Group (USA) Inc.

First printing, May 2011
10 9 8 7

LIBRARY OF CONGRESS CATALOGING-IN-PUBLICATION DATA
has been applied for

ISBN 978-1-592-40644-9

Printed in the United States of America
Set in Apollo MT
Designed by Sabrina Bowers

Contents

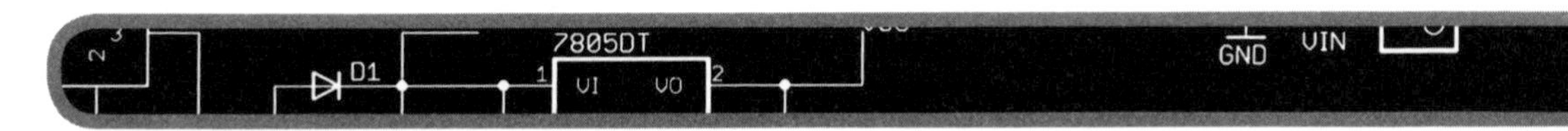

Special Thanks

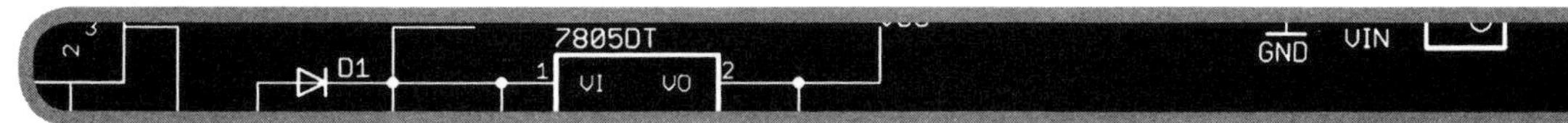

Of course, my specialest (what, I know it's not a word!) thanks go to my wife, Robin, and my sons, Eli and Quinn, who continue to support and enjoy the marvelous ride that Geek Dad has become. And to my parents, Walt and Ellen, my in-laws, Bob and Fran, and my grandmother Charlotte.

Huge thanks go out to each of the folks who sent me project ideas to include in this book. This second book had to come together in a much shorter time frame than the first, and though I wrote or adapted everything you read here, I wouldn't have made it without the wonderful ideas offered up by the awesome and geeky people whose names you'll see peppered throughout.

I continue to be enormously grateful to Megan Thompson at LJK Literary and Jud Laghi at The Laghi Agency for getting this machine running and keeping it on course. And to Lucia Watson, Miriam Rich, and Anne Kosmoski at Gotham/Penguin for helping make Geek Dad such a huge success.

And special nods go to Barbara, Al, and Aaron at McIver's Ace Hardware (I'll always say it MacGyver!) in Fremont, California, who helped a ton with supplies for this book, as well as Paula and Pam at Entourage for being such huge fans.

And lastly to the Saturday morning bowling league parents at Cloverleaf Family Bowl in Fremont, California: Darren and Melisa, Terry and Karen, and Jake and Trav. Our friends and colleagues in this business we call parenthood.

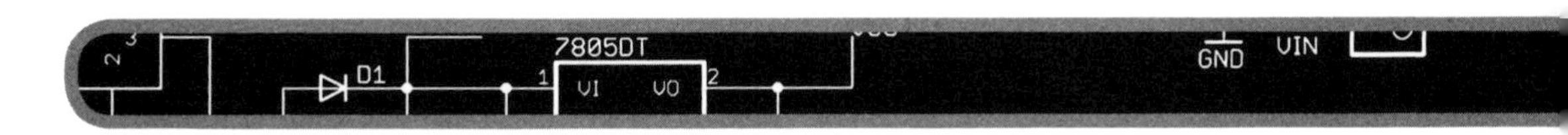

The GEEK DAD'S Guide to Weekend Fun

Introduction

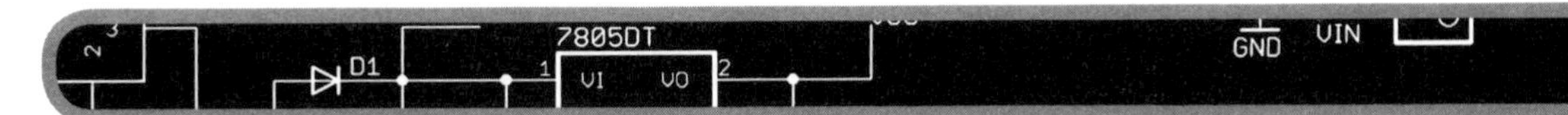

"In reaffirming the greatness of our nation, we understand that greatness is never a given. It must be earned. Our journey has never been one of shortcuts or settling for less. It has not been the path for the fainthearted—for those who prefer leisure over work, or seek only the pleasures of riches and fame. Rather, it has been the risk-takers, the doers, the **makers of things**—some celebrated but more often men and women obscure in their labor, who have carried us up the long, rugged path towards prosperity and freedom."

—PRESIDENT BARACK OBAMA

Geeks like to make things. Indeed, the drive to create is an intrinsic geeky quality (right up there with loving genre fiction and drinking too much Mountain Dew). And while many geeks may not think of themselves as creative in an artistic sense, most geeky pursuits—from rolling up a new D&D character to assembling the LEGO Star Wars Death Star kit (you know, the one with all the cool minifigures)—are acts of creation.

If you are a geek, this creativity comes out in your passion to build things, modify things, or take them apart and put them back together again with new features its designers never dreamed of. And if you are a geek, you have a strong desire to share with your kids this love and imaginative way of looking at the world. But as a parent, you know that passing on anything to our kids (short of male-pattern baldness) can be a tricky proposition.

The challenge is, of course, that because they are kids, they will be suspicious of doing anything you think is "cool." More often than not when Dad or Mom says, "Hey, try this thing I love—it's cool, and you'll really like it!" what kids really hear is "Blah blah blah I want you to do this really lame grown-up thing because it will make me happy to see you bored to death!" But, you know, imagine they hear it in that waah-wahh voice all the adults in the Snoopy cartoons spoke with. So, to really get through to them, what's required is something like reverse psychology, but without all the passive-aggressiveness. You need to let them discover on their own how cool it is to be a geek. My example is this:

As my boys have grown, at various times I've tried to get them interested in my geeky passions with very hit-or-miss success (more miss than hit). They love *Star Wars* in its various incarnations (though, sadly, they're perfectly fine with Episodes 1, 2, and 3 [shudder]), but I could never get them into *Star Trek*. I tried to get them reading superhero comics over and over again, to no avail. *Superman*, *Batman*, the *X-Men*, or *Spider-Man*, nothing would stick; though my younger son loves Simpsons comics, which doesn't quite count. So, when it came time for a major geek transfer-of-passion, a new tactic was required.

I've been a gamer since I was around seven years old. That was the year my geeky uncle Doug gave me the original *Dungeons & Dragons Basic Rules* box set. I remember taking it to school and reading the rules at recess, poring over the adventures and getting really excited by the new world it opened up. I started playing with friends at lunchtime and after school, at first just working through the most basic modules, but by the time I was in high school, we tackled the larger campaigns, running them over multiple weekends. In college, I got the chance to play in original games designed by some very talented GMs (Game Masters), and continued to play occasionally with friends even after getting married.

But becoming a dad took a bit of a bite out of my gaming hobby. So when my kids got old enough (as old as I was when I started), I really wanted to get them into the game so they might enjoy it as much as I have over the years, and it would give us something amazing and imaginative to do together. I was, however, tragically rebuffed the first couple of times I tried to get them excited about the idea. They would politely listen and shrug their shoulders to my "Well, want to give it a try?" and then go back to their video games. I was at an impasse.

The great breakthrough came when I ordered the new D&D 4.0 Player's Handbook. It came from Amazon, and I spent a couple days checking through the new rules, admiring the streamlined system.

And then—this is key—I just left the handbook out on a table in plain sight.

A day or two afterward, my older son walked up to me while I was working on the computer, the book in hand. "Dad," he said, "this looks pretty cool. Can we try it sometime?"

I took a deep breath, and strenuously worked to hide my excitement. "Well, sure, that'd be cool. Why don't you go ahead and read through the rules a bit, and then we can talk about what you might want to try."

"Okay," he said.

And that was that. Not only did he start hounding me to play, he got his younger brother and their best friend from the neighborhood excited as well. Eventually, we created characters for them, and I started building a simple starter adventure to get them introduced to the idea of a pen-and-paper role-playing game, or RPG (and get myself more familiar with the new rules).

What I took away from that experience is that, as parents, we can try to force cool things on our kids and then hope they will get as caught up in them as we are. But in the end, that strategy doesn't work. It's far, far better to let them discover these things themselves. (I'm not saying you can't occasionally leave a little bait lying around, though!)

WORKING FOR THE WEEKEND

Back in the fictional *Leave It to Beaver* 1950s, Dad was a mostly benign but distant figure who came home late, ruled the dinner table, and then relaxed with the paper while Mom took care of getting the kids to bed (and pretty much everything else related to running the household). Maybe there was a ball tossed about on the front yard once in a while, but not much more than that.

Thankfully, we, as modern geeky parents, do things a little differently. First, Mom and Dad both work, and household management must be shared. But still, playtime is scarce. The little time we have after getting home from work is taken up by our kids' homework and our own responsibilities. Not a lot gets done together as a family. Indeed, it's rather amazing we have any time at all, when you think about the demands put upon us. Consider the average weekday:

Sleep: 8 hours
Work (w/ lunch hour & commute): 10 hours
Morning toiletries and breakfast: 1 hour
Dinner prep, consumption, and cleanup: 1 hour
Prep for bed, wind down: 0.5 hours

That's 20.5 hours a day reserved already, leaving a paltry 3.5 hours for "other." But there are bills to pay, laundry to be done, homework to be reviewed, errands to be run, and more. In short, we're lucky if we get to spend any time with our kids on the weekdays.

That leaves us the weekends. The sacred days. For some of us, the only days of the week when we actually get those mythical eight hours of sleep. And they are the days we actually get to spend serious time with our kids.

The point of the projects in this book is that they all easily fit into the spans of time you can carve out of your weekends to do things with your kids. Of course, it doesn't hurt if you strategize a bit by getting up early Saturday morning to get the weekly shopping done when there are no lines, or mowing the lawn, tidying up the garage, and fixing the latch on the gate that hasn't been working quite right, before the kids get up. If you do the stuff you have to do, you can set aside some free hours to do the things you want to do: like build weird and awesome things with your kids. The projects in this book are worth making time for, and needless to say, you and your kids are worth it, too!

HACK IS NOT A DIRTY WORD

A word about the concept of "hacking." Many of the projects in this book involve hacking existing objects to create something fresh and unique. This may make some people a bit nervous since the act of hacking has gotten a bad reputation. This may well be because of the way it's been used at times in connection with illegal activities. But at its core, that's not what this activity is about. The true definition of hacking in our modern culture is to repurpose something for your own use (which may be significantly different from the use originally intended by the creator of the thing). While the more salacious connotation for this is to take control of electronic security and computer systems in order to access data or control devices—that is, repurposing them illicitly—there is a very strong movement around the world that thinks of hacking in a much purer way. This DIY movement celebrates adding creativity on top of creativity to build new and interesting things.

If you own something, it's your possession. You may keep it with you and use it at your discretion, for your own purposes. You control it. But if you don't know how it works, do you really exert complete ownership over it? If it started behaving in a way that you didn't understand and couldn't change, would you no longer be able to control it? Would you still truly own it?

The modern hacking community has asked this question and come back with an answer: "If you can't understand how something works, you don't truly own it." This is a powerful idea and one that changes the way we see the world around us. By hacking objects with your kids, you open their minds and build their confidence through a new, hands-on way of learning.

And, once your kids feel comfortable opening things to see how they work, they can get to the place where they can fix and even repurpose them. It's a skill and a sense of mastery that will last them a lifetime.

WHAT'S INCLUDED IN THIS BOOK

This is a book of projects, yes, but they aren't meant to be rigid paint-by-numbers instructions (or at least not JUST that). These are all meant to be ideas for you and your kids to explore together. It's very likely you don't have all the same spare parts lying around your house as I do. You probably have different LEGO sets and tools and so forth, so if you don't have the exact materials called for in a project, don't fret. Look in your garage, your basement, your craft room and see if you can find something else that might work.

That's hacking. That's adapting. That's (now don't be scared) *engineering*!

And that's my point: Don't treat these projects like the recipes for delicately crafted soufflés that may implode if you deviate from the instructions slightly. Rather, think of them as guidelines that can get you from a beginning to an end. But it's up to you and your kids to figure out how best to apply the guidelines. Indeed, depending upon your available materials, your know-how, and just plain luck, you may find a better way of doing some of these projects!

You'll also find, peppered throughout this book, a number of "celebrity" projects and anecdotes. I put that word in quotation marks because who you consider a celebrity may differ from my take on the title. David Hewlett of *Stargate: Atlantis* fame certainly has his measure of fame, as does Patrick Norton of *Tekzilla*. Chris Anderson, the founder of www.geekdad.com and the editor in chief of *Wired* magazine, has pretty good name recognition, as does Ken Jennings, the superchampion of the game show *Jeopardy!*. Other names may not be quite as well known. But, while you may not recognize everyone who has earned the title "celebrity" in this book, know that they are cool folks who have earned some cred in the world of geeks.

PROJECT INFORMATION

At the start of each project, you'll see a table with summary information to give you an idea what to expect from it, and there are some symbols not unlike what you see in a restaurant or hotel review to explain cost and difficulty. Here's a legend to explain their meaning.

PROJECT	TITLE OF THE PROJECT
CONCEPT	A quick overview of the project so you can decide if it's of interest to you
COST	$ = $0 to $25 $$ = $25 to $50 $$$ = $50 to $100 $$$$ = $100 on up In many cases I'll exclude the cost of tools and materials that are so common that you probably already own them or could find or borrow them easily.
DIFFICULTY	= easy for primary school–age kids to grasp and enjoy = for secondary school age and up = for junior high and up = high school age There is a wide variety to these projects, from the very simple to the rather complex. Some of them can even be adjusted to be more or less in-depth. It's up to you to gauge your kids' ability level and attention spans, and pick the right projects to share with them. After all, you know your kids better than I!
DURATION	= 0–15 minutes = 15 minutes to 1 hour = 1 to 3 hours = 3 hours and longer While the journey is often just as important as the destination, it's nice to get there, too. You and your child can finish most of these projects in an afternoon or evening of cooperative work.
REUSABILITY	= one time only = reuse once or twice = multiple reuse possible = good forever Hopefully, you'll find that most of the projects in this book will have serious repeat value.
TOOLS & MATERIALS	A list of the basics required to perform the project. I'll do my best to suggest how to make do without buying too much.

One thing you'll notice as you go through the projects in this book is that they don't often hit the extremes (superlong, superdifficult, or superexpensive), save reusability. This book isn't about long, costly, or involved projects that take too much work before paying off in the fun department.

Indeed, my goal with this book (as with the last) is to provide a resource for inexpensive and accessible projects that are just a bit outside the norm. Hopefully, each one has a touch of science in it, too, so that when you finish the project, you can look back and say "Hey, now I understand how that works!"

But more important, I want you and your kids to have FUN. To spend TIME together learning, building, and playing. I want you to suck them into the geeky things you enjoyed as a kid, which made you the geeky parent you are today. These projects should be just an excuse to have some wonderful shared experiences that you'll remember for a long time. As a bonus, hopefully, they will activate the geeky genes in your kids.

So I bid you to go forth with your kids and build something together!

HACK IT, BUILD IT, PLAY IT

NERF Dart Blowgun

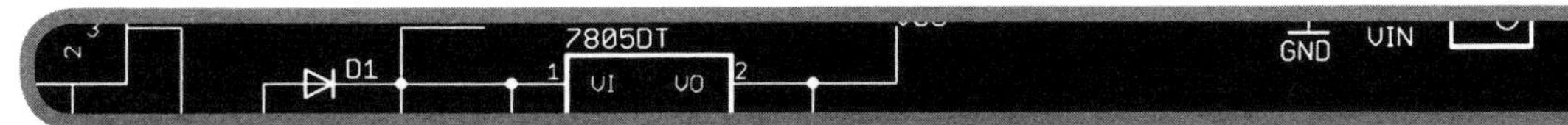

Two slightly interesting facts about NERF: first, it's as old as I am (not telling), and it's an acronym that stands for "Non-Expanding Recreational Foam." There, you've learned something you didn't know before you read this chapter.

But seriously, if your kids are like mine, there are NERF blasters in your home, and the ubiquitous darts ALL OVER EVERYWHERE. You can likely move a few couch cushions or the doggie beds and find a couple lost darts.

I'm glad NERF blasters exist. When I was a kid, I had a pellet gun that looked one heck of a lot like a .45 caliber automatic. These days, my far more protective parent instincts tell me that was CRAZY! Those fluorescent orange pseudo-guns with foam darts are a much better way for kids to get out their gun play.

Thankfully, this book isn't going to debate the finer points of overprotective parenting. What it is going to do is give you a project in which you can build an alternative to the store-bought NERF guns that'll launch the darts FARTHER and HARDER. Your choice will be whether you share this project with your kids or keep it to yourself for a while.

PROJECT	NERF DART BLOWGUN
CONCEPT	Make a blowgun for NERF darts that will launch them much farther than store-bought guns.
COST	$-$$
DIFFICULTY	⚙ — ⚙⚙
DURATION	☼ — ☼☼☼ (1 to build, 3 to play with)
REUSABILITY	🌐🌐🌐 — 🌐🌐🌐🌐
TOOLS & MATERIALS	NERF darts, ½ inch copper pipe, ½ inch copper fittings (as needed), silicone sealant

NERF advertises its Longstrike blaster (designed to look and act like a sniper rifle) as firing a dart "up to 35 feet," and it costs around $30. I'm not trying to compete with Hasbro! The NERF blasters are great toys—I can only dream of replicating their clip loading systems. And their Vulcan (a replica of the M60 machine gun) is just an awesome toy (props to www.makezine.com contributor John Parks, who hacked together an autofire turret for the Vulcan). We wouldn't have all these spare darts lying around if our kids didn't love them in the first place.

However, this project provides a way to have even more fun with them—to hack them, if you will. And maybe teach our kids a little science in the process.

If you take a NERF blaster apart, you'll learn that the propulsion system is based on a mechanical force applied to the dart, usually by a spring-loaded firing pin. A dart is loaded, the pin is cocked, the trigger pulled, and the pin hits the dart and launches it down a barrel. This works pretty well. But it can work better.

Aside from mechanical force, the other good way to propel something down a tube is by pneumatics. Indeed, if you consider it, the way real guns work is via pneumatics. The explosion of power causes rapidly expanding gases to launch the bullet down a rifled barrel at high velocity. So why can't we replicate that with NERF, which is basically a scale model of a "real" firearm?

A blowgun does just that. The projectile is put in the barrel, and the human supplies the rapidly expanding gas (exhaled air) to propel it. What's truly fascinating is that, for the NERF darts, this system actually works better than the mechanical one.

The most basic part of this project can be done in one step: Go to your local hardware store and pick up a 20–24-inch section of ½-inch copper pipe. This is the kind of pipe that is commonly used for water in domestic plumbing. A piece like that will cost about $1.

Take a NERF dart, insert it into one end of the pipe, with the "business" end of the dart pointed up the length of the pipe. Hold the pipe up to your face and aim it at some unsuspecting target. Put your mouth on the pipe, take a deep breath, and, in as short and sharp an effort as possible, blow.

Without much practice, the most basic blowgun design I just described can hit targets more than 60 feet away. That's longer than 35 feet!

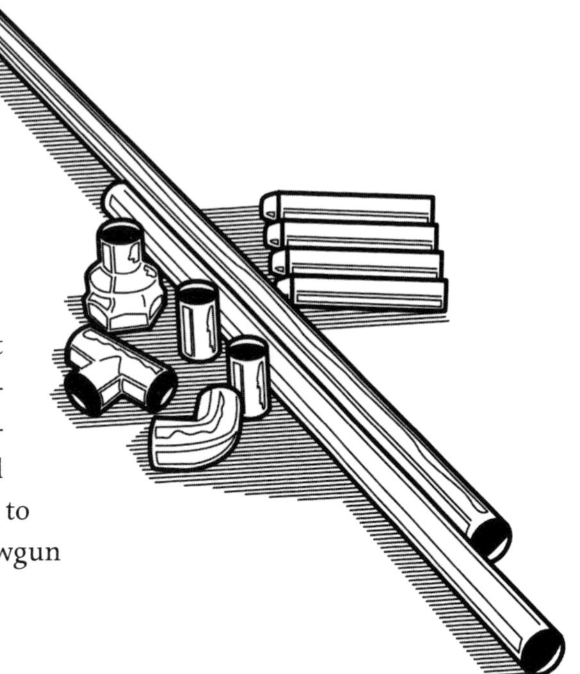

But one length of pipe does not a project make, so let's try upping the complexity of this idea. If you go down to your local hardware store and start browsing the fittings that go with the ½-inch copper pipe I mentioned above, you'll find 90-degree elbows, 45-degree turns, "tees," and reducers. Here are some ideas as to how to use these extra parts to make your blowgun even cooler:

REDUCER: As the name suggests, a reducer is a fitting used to reduce from one pipe size to another. I picked up a $\frac{3}{4}$-inch-to-$\frac{1}{2}$-inch reducer and found out that it works as a perfect mouthpiece for the blowgun. Using the reducer on the blowing end of your blowgun allows you to purse your lips and blow into the gun like a trumpet rather than wrapping your lips around the end of the blowgun. This is more sanitary, and it lets you build up air pressure for a nice burst more easily than without the mouthpiece.

ELBOWS: Taking a shot from cover can't be any easier if you can literally shoot around corners! Link two lengths of pipe with an elbow, load a dart into the end tube, stand at a corner with the end tube directed around the corner, and shoot!

TEES: Tees allow you to branch off multiple pipes from one blowpipe, so you can rain down destruction with a multishooting blowgun. Just don't get too carried away, because you or your geeklet's lung capacity will be the ultimate arbiter of how many darts you can shoot, and how far.

One thing to understand when working with the fittings: The inside dimension of the fittings is only larger than the outside dimension of your pipe, meaning the pipe is meant to slip inside the fittings, but fittings won't slide into each other. The way to attach together a series of fittings is to cut small 1-inch-to-1$\frac{1}{4}$-inch lengths of pipe to use as connectors between the fittings.

As you put pieces together, you're going to want them to actually stay together and not flop around. Rather than soldering the pipe, as is usually done for true plumbing, use a clear silicone caulk as you might to seal around fixtures in your bath or kitchen. It'll keep the fittings connected and pretty airtight, and if you end up deciding to take things apart again, it's not that hard to break the seals. Just put a little bead of the caulk around the surface of the pipe piece that's going into the fitting, and slip it in. Once you have your pieces together, let them sit and dry for a couple hours (depending upon the instructions on the caulk packaging).

Because of the versatility of the fittings, you could even put together a series of pipe constructs that are interchangeable, depending on need. Maybe you'll have a mouthpiece and short pipe section that can be used as a simple single-shot device, but also carry an elbow attached to another short length, which your NERF warrior can slip on the end for the emergency corner shot. And have a special multibarrel piece to add when the extra firepower is needed.

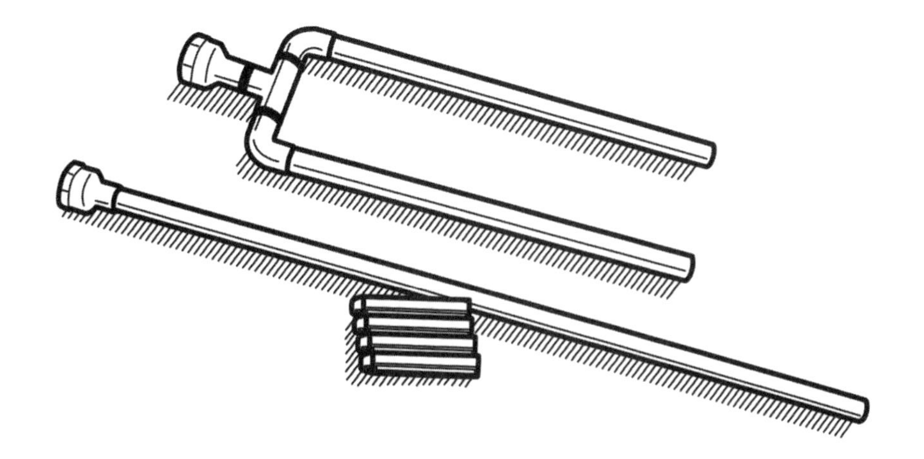

After you've built your arsenal, it's time to play. Maybe try some target practice and see what the accurate range of the blowguns is versus the traditional NERF blasters. Or just go have a backyard firefight and see how well the stealth blowgun stacks up. Maybe your geeklet will become a NERF ninja!

Twenty-First-Century Superhero Cape

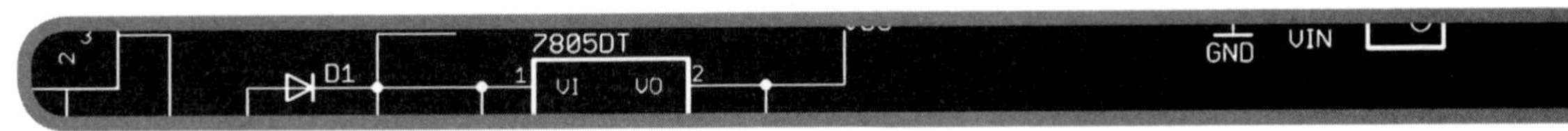

(Project Idea by Jenny Williams)

Look! Up on the bed! It's a bird! It's a plane! It's your geek kid!

What could be more basic to a geek child's wardrobe than a cape? How many of us tied blankets, tablecloths, or towels around our necks and ran around with our arms up in the air, making flying noises, pretending to be our favorite superhero? And what could be more empowering for your supergeek-in-training than creating his or her own cape? This is a fun project to do together. It's passing on a geek legacy, like Bruce Wayne working with Terry McGinnis in *Batman Beyond*!

And even better, why be restricted to just one superhero identity? The nature of a cape lends itself to being lined or reversible, so I'm going to show you how you and your young ward can make one cape that will satisfy the need to be two different heroes.

PROJECT	TWENTY-FIRST-CENTURY SUPERHERO CAPE
CONCEPT	Sew a reversible superhero cape for all of your saving-the-world needs.
COST	$–$$ (depends on the fabric)
DIFFICULTY	— (For the making; 3+ for the wearing, your child must be careful to avoid any potential choking hazards)
DURATION	—
REUSABILITY	—
TOOLS & MATERIALS	1+ yards each of two different fabrics that are of similar weight and consistency; 1½ to 2 yards of 1½-inch (or wider) ribbon (or silk rope, snaps, or other fastening); matching thread; measuring tape; iron and ironing board; scissors; pins (sewing method); sewing machine (sewing method); iron-on hem tape (non-sewing method)

One caveat before starting this project: The most desirable way of putting this together is by sewing, either by hand or with a sewing machine. Don't let this scare you! If you are a Geek Dad or Mom with zero sewing experience, this is the place to start, for it involves about the simplest straight-line stitching you're ever going to see. I do suggest borrowing time on someone's sewing machine to make it go faster (and maybe the machine owner can even give you some tips).

Alternately, there is a no-sew option for this project. It won't be quite as durable, but it will be more accessible for those with needle issues.

STEP 1: Fabric Selection

Choosing one's superhero cape fabric is an intensely personal experience, and while it may be shared with a loved one, the final decisions should always be left up to the superheroes themselves. Meaning, let your geek kid choose his own colors/patterns!

Of course, that doesn't mean you can't suggest certain types of fabric that'll make the process easier. For example, stretchy or shiny fabrics are harder to evenly cut and to sew, and it's usually easier to pair two of the same types of fabric (cottons, blends, wools) so that the cape sweeps well, particularly during the all-important turn-and-dash move. And bright solids will always lend a bolder, more "superheroic" air.

You'll need at least 1½ yards of each fabric if you want a wide flair to your cape (or if you are making it for a grown-up!). If you purchase a fabric that is likely to shrink significantly, increase the yardage by at least a half. A trip to your local fabric/crafts store will offer a tremendous number of choices in style and cost. But don't be shy about hitting the discount bin for remainders. Fabrics aren't sent there because they're bad in some way, just that a bolt (the unit role of fabric) may have been sold down to the last yard or two, which often isn't enough for a normal pattern. You may be able to find an awesome bargain. Of course, you may already have some great fabrics at home to work with, and can save yourself the trip!

STEP 2: Fabric Preparation

This cape has the potential to become a prized garment, worn on adventures all through the universe. Your budding superhero will likely face foes from the mundane to the cosmic, and will have to get dirty sometimes saving the world.

Which means you'll want to be able to wash the thing from time to time.

Knowing that, you'll need to run the fabric through the washer and dryer once before doing anything else to it. All fabrics will adjust in some way when they go through a wash/dry cycle. Make sure you know what cycle is appropriate for the fabrics you have chosen (another good reason to keep the choices similar in material—so you can get away with one wash load before assembling; just don't run white and red through together on a hot-water cycle, or your new hero will be The Pink Avenger).

You can even use your imagination to make this part of the process magical for your superhero-in-training: Explain that you're not just prewashing the fabric, you're putting it through a special high-tech/magic process to imbue it with special qualities!

Once the fabrics come out of the dryer, cut off any loose strings or frayed edges and iron out any wrinkles (also known as "tempering the fabric"!).

STEP 3: Measurement and Cutting

Now you need to decide what kind of cape your superhero wants. Will it be a half cape, falling just below the waist à la Robin in the 60s or many stylish female superheroes? Or should it be down to the knee, which adds a rakish look and makes for a really good sweeping swoosh when one turns and rushes off to action? Or should the cape be long enough to just scrape the floor like the Dark Knight's, hiding all kinds of gadgets and allowing for full-body protection? Decide with your child on the appropriate length and measure him or her accordingly from neck to waist, knee, or floor and add an inch (I'll explain why you add an inch in a moment).

Width will be important as well. Should it hang just off the shoulders or be wide enough to wrap around the body? The width measurement should be for the widest part of the cape—usually at the bottom. Add one inch to that measurement as well.

Then, with these two measurements, cut a rectangle out of each fabric.

STEP 4: **Shaping and Preparation for Sewing**

Put the two pieces of fabric together, with all corners and edges matching. The "faces" of the two fabrics (meaning the sides that will eventually be the outward surfaces of the cape) should be facing each other now. This may not be an issue if the fabrics don't have printed and unprinted sides, but if they do, the unprinted sides should be facing out right now.

If you're happy with just a rectangular cape, you can skip to Step 5. But if you want a little more shape, like a flare to the cape (wider at the bottom than the top), try this: Fold the matched fabrics in half lengthwise and cut the long edges in a straight line from where the bottom corners meet, to a spot a few inches in from the top corners. Unfold, and your fabric will now be in the shape of a trapezoid.

STEP 5: **Sewing the Cape**

Here's where you need to choose: To sew or not to sew—that is the question! For seamsters, keep reading. For folks who want the easy way out, skip down to the heading "Alternate Method."

If you're going to sew, you and your little geek (depending upon age, of course) need to pin the edges—meaning you take sewing pins and insert them perpendicular to the fabric edges, once down through and up again, every six inches or so, until the entire perimeter of the fabrics is pinned:

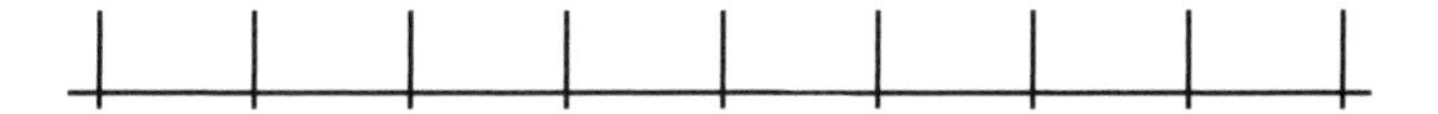

All this does is ensure the fabrics don't slip against each other and misalign while you sew.

With your sewing machine (or, much more slowly, by hand), sew a seam $\frac{1}{2}$ inch in from the edge along both sides and the bottom of the cape, leaving the top of the cape open. This is a great place to introduce your child to sewing (if he's not already well informed), because the workings of a sewing machine are just really, really cool!

Clip the bottom corners off at 45-degree angles without cutting through the seam. Once you've done this, skip down to Step 7.

ALTERNATE METHOD:

Okay, so sewing isn't quite right for you, either because you don't have access to a sewing machine or you want to tackle this a little more quickly. No harm, no foul! Instead, we'll using iron-on hemming tape. Basically this is a heat-activated two-sided adhesive tape for fabric.

Cut lengths of the tape equal to each edge of your cape. On each side and the bottom of the cape, carefully fold over the edge of the top layer of fabric, and put the tape on the other fabric so it follows in the inside line of the cape. Then fold the top fabric back down so it covers the tape, making a fabric-and-tape sandwich.

Once you have the tape installed on all three sides, pin the fabric just like in the sewing method above so nothing will slip. And heat up your iron to the temperature recommended on the hemming tape packaging.

When the iron is hot, carefully move the cape to your ironing board, and follow the instructions on the hemming tape packaging to adhere all three sides of the fabrics together.

STEP 7: Right Your Cape

Now you need to right the cape. Turn the cape right side out again by stuffing everything through the hole you left when you didn't seal up the "top" of the cape. Use a pointy tool like a capped pen to make sure the bottom corners are sharp, poking the edges from the inside if needed.

No matter which of the above methods you used, you should now iron the heck out of the cape to get a nice crisp edge all the way around.

STEP 8: Pleating (if needed)

Now's the time for the first test-fitting. Hold the cape up to your superhero and wrap the top around his or her neck and shoulders in the way he'd like to wear it. If the top corners overlap too much, you can put in some pleats (extra folds in the fabric) to make the cape drape correctly. To do this, measure how much the corners overlap and subtract that amount from the total width of the top edge of the cape.

Add pleats symmetrically along the top edge, pinning them as you go (with hemming material in place if that's your method), until the top measurement equals the number you came up with. Example: 29-inch-wide top minus 7-inch overlap when wrapped around superhero = 22-inch-wide final top measurement after pleats. Several small pleats are usually better than one or two big pleats. Sew and/or iron to make the pleats permanent and get just the shape that fits.

STEP 9: Adding the Cape Tie

The kind of tie used to clasp the cape at the neck is almost as important, stylistically, as the cape itself. A ribbon is quite handy, and that's what I'll describe using here, but you have a number of other options, including snaps, eye hooks, grommets and silk rope, and so on.

Determine how long a ribbon you will need by adding at least 24 inches to your final cape-top measurement. The longer the ribbon, the easier it will be to tie, but the more it will hang down in front.

Cut the ribbon to the desired length. Locate the middle of the ribbon and the middle of the cape top. Here again we have the sew versus no-sew option. If you are sewing, fold the ribbon middle over the middle of the cape's raw edge. Pin in place. Pin outward from there, folding the ribbon over the top. Make sure that you fold the ribbon evenly over the front and back of the cape. Pin all the way to both ends of the ribbon, tucking in the raw ribbon edges at the end. Sew across from one end of ribbon to the other, close to the ribbon edge so that the ribbon now edges the neck of the cape and creates a tie for the cape.

If you are not sewing, cut a piece of iron-on hemming material to match the ribbon, and fold the tape and ribbon together (with the hemming tape inside) over the middle of the cape's raw edge. Pin them in place at the center and pin outward from there, folding the hemming material and ribbon over the top. Make sure that you fold them evenly over the front and back of the cape. Pin all the way to both ends of the ribbon/hemming tape, tucking in the raw edges at the end. Iron it all together so the ribbon now edges the neck of the cape and forms a tie rope with which to fasten the cape.

STEP 10: The Final Step

Put the cape on your superhero and send him or her out to save the world!

WAYS TO WEAR A CAPE:

- The cape can be reversed to suit your child's mood or superhero tasks. When you pick the fabrics for either side, you could even deem that one side is the hero side, and the other is the villain side!
- Turn the cape so that it is all over one shoulder, leaving the other arm free for swordplay with dastardly foes.
- Wear it centered, pulling it around the wearer to stay warm.
- Wear it centered, but fold each side back in warm weather, exposing the contrasting underside.

Even Geekier Ideas

UNIQUE DESIGNS You can embellish this design with many extra, added features. Add a star or other superhero sigil cut out of contrasting fabric. Or maybe loops and pockets for special gear. You can sew on (or iron on, using hemming tape) embellishments after cutting out the fabric shapes and before pinning right sides together.

LIGHTS What about lights? A fun add-on might be to sew on a string of battery-powered LED lights around the edges of the cape (But note: This will mean you can't wash it). Or use some glow-in-the-dark fabrics to make for an amazing nighttime superhero cape!

ALTERNATIVE USES If you make this cape wide enough, short enough, without pleats, and with a long-enough ribbon, it can also serve as a wraparound skirt for supergirls, or a superkilt for boys!

Trebuchet with LEGO Bricks

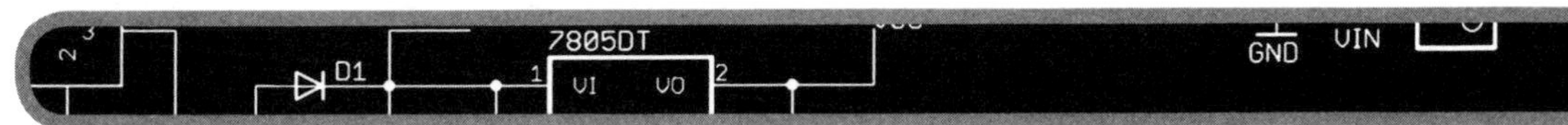

Ah, siege warfare! Who amongst us hasn't dreamed of rolling up a few massive medieval mechanical weapons to the perimeter of an enemy castle and letting loose with a few rounds of flaming projectiles?

Okay, well, that may just be my family after a filmfest of all three extended-editions of the Lord of the Rings movies in one sitting. But launching destructive objects long distances is usually a good way to engage kids. And the best part? You may well have all the parts you need to pull this model project off just lying around.

PROJECT	TREBUCHET WITH LEGO BRICKS
CONCEPT	Build a scale-model siege engine out of LEGO parts.
COST	$–$$$
DIFFICULTY	⚙⚙ – ⚙⚙⚙
DURATION	☼☼☼
REUSABILITY	🌐🌐🌐 – 🌐🌐🌐🌐
TOOLS & MATERIALS	LEGO bricks; Mindstorm NXT or Technics building pieces; string; towel or other fabric; 100 ¼-by-1½-inch fender washers; key ring

A trebuchet is a great big sling with a little extra engineering, and there are versions of it recorded back to the fifth century. The design we're working from, known as a counterweight trebuchet, probably started wreaking havoc on walled cities in the twelfth century. It remains well loved even today amongst geeks who like to re-create the technology of earlier ages. And it's a terrific way to teach your kids some basic scientific principles.

The way a trebuchet works is pretty simple. While a classic catapult uses the energy stored up by bending wet wood to launch things, the trebuchet uses the principle of the lever. A heavy weight is suspended from one end of a long lever. The axis of the lever is set a short way from where the weight is mounted. A sling construct is attached to the other end of the lever. A projectile—equivalent to perhaps 0.5–1.0 percent of the weight of the counterweight is loaded into the sling, and then the whole setup is wound up so the weight is raised high, the lever is pushed low, and the projectile in the sling sits nearly under the axis. When the weight is released to fall, the lever arm rotates around the axis, the long end shoots up, pulling the sling around in an arc, and at the optimum angle, one end of the sling releases, letting the projectile fly.

In their life-size incarnations, these machines could fling projectiles weighing more than a ton. For this project, we're going to set our sights quite a bit lower.

You can make trebuchets out of a wide variety of materials. Indeed, there are kits and downloadable instructions online for wooden models that are very good simulations of a medieval trebuchet. But for simplicity and customizability, I chose to do this with LEGO bricks.

I'll admit I cheated a little, though. I didn't use just plain old LEGO bricks. I have a LEGO Mindstorms NXT kit, which is a programmable robot-building system with pieces that are more useful if you are making things with a mechanical purpose. If you don't have a NXT kit, many of the Technics kits have similar pieces. Or you could try to do this project using Erector or VEX pieces as well. Once you know how a trebuchet works, you can use many kinds of materials to build it.

STEP 1: Build the Lever Arm

The lever arm needs to be a pretty rigid construct. You can use the beam pieces that come with the NXT set, along with the joining brackets, to achieve this. Regular LEGO bricks won't work, because the torsion that will be applied by the counterweight will overcome the way bricks lock together. What you need are pieces that will lock rigidly.

The length of your lever should be about $\frac{1}{4}$ for the counterweight and $\frac{3}{4}$ for the throwing arm. There must be a way to attach the lever arm to your superstructure at this $\frac{1}{4}$ and $\frac{3}{4}$ fulcrum point that will allow it to swing freely, like an offset seesaw.

For the counterweight, I put two simple crosspieces between the beams and threaded enough stamped washers to fill all the way

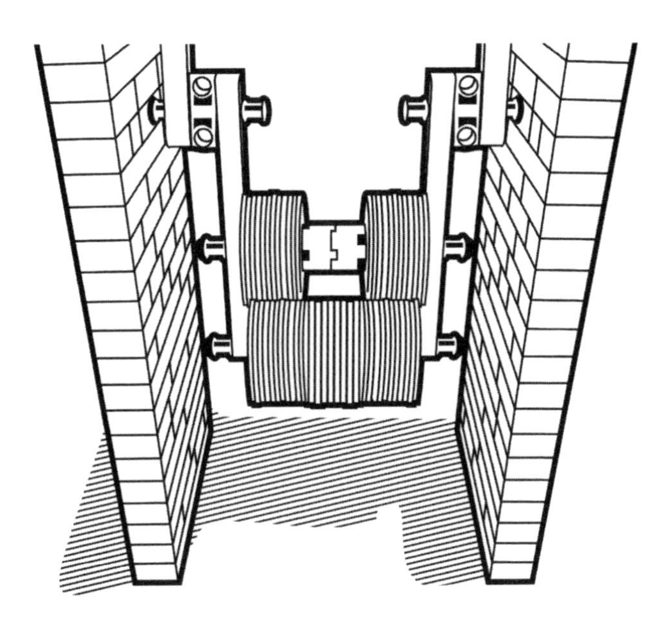

across (at least 20 each beam for 40 total). The whole idea is to get a pretty big weight on the short side of the fulcrum, so instead of washers, if you have something like lead fishing weights, which will be even heavier for the space they'll fill, that's a good upgrade. Actual numbers will depend upon the scale of the trebuchet you build, but the ratio of your counterweight to the weight of your projectiles should be around 10 to 1, or more.

STEP 2: Build the Superstructure

Here's where the traditional LEGO bricks come in most handy. All the superstructure has to do is support the weight of the lever arm and counterweights at rest, and be able to deal with the rotational forces when it's in motion.

Use a variety of 2x2, 2x3, 2x4, 2x8, and even 2x10 bricks to build the structure walls for supporting each side of your lever. Make each structure wall 2 studs wide by 12 studs long so you can maximize the combinations of bricks available. The most important thing is to stagger seams. If seams line up on multiple levels, they weaken the overall structure. If you can, build the walls on a LEGO base plate. Separate them by the width of your lever arm construct.

The height of the support walls depends on the length from the axis point of your lever arm to the outside edge of your counterweight. Since the counterweight is going to swing through an arc, the axis point has to be high enough so the low point of the swing will still keep the counterweight off the ground.

To attach the lever arm to the structure walls in this design, use a Technic-style brick with holes in it, through which a crossbar can run. This allows the lever arm to rotate freely.

An additionally helpful design feature for the support walls is a crossbeam near the stop, to help keep the walls vertically true and to act as a stop for the lever arm when it finishes its rotation to launch of the projectile.

STEP 3: **Build the Sling**

The sling is the heart of the trebuchet. It's the idea originally created to extend the throwing ability of the human arm. On a trebuchet, the lever replaces the arm, but the sling remains the same.

There are three components to the sling: the pouch, where the projectile rests; the cords; and the thimble, which is the release that allows the trebuchet to actually launch the projectile.

For the pouch, I just cut a rectangle about 2 inches by 4 inches out of a shop towel, and then trimmed the corners off. At the ends, I punched through holes so I could tie them off to the cords.

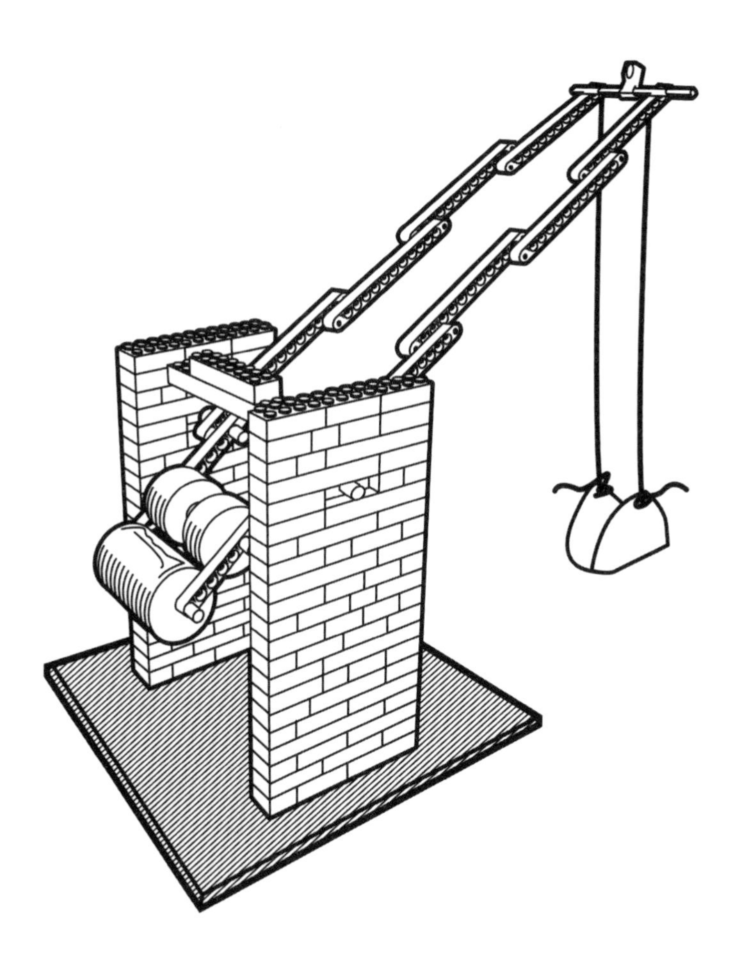

There are two cords. The cords should be made of string for this build. Each one should be about the length of the long section of the lever from the axis to the end. If you ever decide to build on a larger scale, you could use lightweight rope instead.

One end of each cord ties off to an end of the pouch. One cord then ties off to the end of the long part of the lever. The other cord's end gets tied off to the key ring, which will act as the catch-hook which keeps the whole launch package swinging around until the top of the arc. Both cords should still be of roughly equal length, so that when a projectile is loaded in the pouch, it will hang from the lever arm evenly.

STEP 4: Fetchez la Vache

Now it's time to have some fun launching projectiles and seeing how far they can go. Remember, you may need to do a bit of fine-tuning to get everything running smoothly—your projectiles should be a very small percentage of your counterweight. I used the small colored balls that come with the Mindstorms NXT set, but you could easily use standard 2x2 LEGO bricks.

To launch, hook the key ring end of your sling over the catch at the top of the lever arm. You load your projectile into the sling, and then pull downward and inward under the fulcrum, so you are pivoting the counterweight upward on the opposite end. When you let go, the counterweight will pull down, pivoting the arm, which draws the sling out, around, and over the top.

When the sling flies over the top, the key ring slips off the catch, allowing the projectile to escape and launch into the air. Your sling has to allow the key ring end to slip off the catch at the optimal part of the lever rotation so the projectile launches at a roughly 45-degree angle to the ground. So tweaking each part of your build can maximize the effects. But it's the perfect way to learn from trial and error.

And learning's half the battle.

Hack Your Own Sound Box

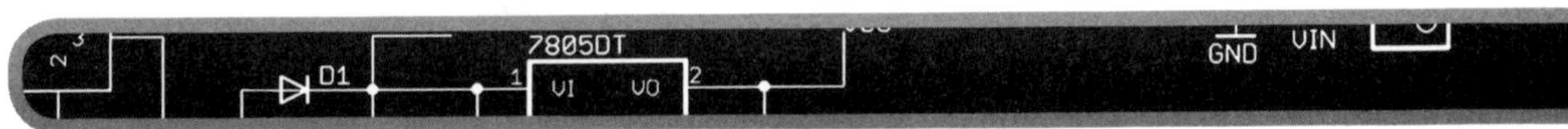

(Project Idea by Dave Giancaspro)

As I explained in the Introduction, hacking is not evil. Hacking is just taking ownership of something and applying your own experience and know-how to turn it to your purposes. If your goals are malign, then that can be bad, but if your goals are to learn, be creative, and have fun yourself and with your kids, then more power to you!

The things we can hack are all around us. We just need to look closely to see them. How about the toys that come with your fast-food meals? Probably some fun to be had there! Or that old VCR in the garage? You could crack that puppy and learn a lot! Or how about that wacky greeting card you received from Aunt Petunia last Arbor Day?

Huh?

Yup. The technology going into greeting cards these days is actually pretty sophisticated. And we're going to have some fun with it in this project.

PROJECT	HACK YOUR OWN SOUND BOX
CONCEPT	Transplant the electronics from some talking greeting cards into a custom enclosure, add your own controls, and make your very own digital audio recorder
COST	$$
DIFFICULTY	
DURATION	
REUSABILITY	
TOOLS & MATERIALS	4 Dinotalk's Talking Paper kits from www.dinotalk.com; AAA battery holder Radio Shack 270-412; 1 SPST switch for power Radio Shack 275-1565; 4 SPST toggle switches Record Enable Radio Shack 275-0624; 5 SPST momentary push-button switches; 2 packs Radio Shack 275-1547; Red LED Radio Shack 276-0307; LED holders Radio Shack 276-0079; 100 Ohm resistor Radio Shack 271-1311; 1 Project Box to stuff everything in, Radio Shack 270-1803; stranded wire red and black plus at least one other color

This is perhaps the most challenging project in this book, since it has electronics that are more complex than the ray gun (though on par with the LED lanterns), and you are actually fabricating your own case by making a template, drilling holes for the component placement, and doing a lot of soldering of wires all over the place. But as with other electronics-based projects, the idea is simple, and if you think about controlling the flow of electricity, everything you'll do here makes sense.

The simple idea is that we're going to take the guts from four greeting cards that have audio record and playback capabilities (they can usually store 5–10 seconds of sound), rip them out, rewire them with some much cooler buttons and switches, and mount the whole thing into a cool-looking box, resulting in a digital recording and playback device. We're using blank cards from a specific source rather than ones you'd go down to your local card store and pick up, because we need to make sure everything looks the same. You can always try this with over-the-counter cards—the guts should be similar, but for the sake of making the project description clear, we're doing it this way.

Each card has some very specific components: power, a power switch (on/off actuator), a microphone, a speaker, a RECORD button, a PLAY button, and the actual controller module—a small circuit board into which everything is wired. Since we're going to combine the guts from four cards for our device, we can consolidate some of the parts. We'll need only one power supply, one speaker, and one microphone. But we will need all four controller modules. We're going to take all those parts and add a few things of our own—individual switches and buttons to allow us to pick which of the four modules records when we hit the RECORD button. We'll also add a better power supply and a fancier RECORD button, as well as a cooler ON/OFF switch. We'll wire it all together, plop them in a project enclosure box that we'll customize for all the pieces, and in the end, it'll look like a wicked gadget Q might have built for 007.

PREPARING THE BOX

STEP 1: You'll start by drilling out the lid and side of the box to accommodate the switches, the indicator light (LED), the speaker, and the microphone. (Note: this project is based on using the specific project box listed in the materials, but you can improvise with whatever you have handy.) This is one of those "it's completely up to you how you make this look" deals, but obviously I won't leave you hanging for a place to start. The sketch of my drill pattern for the box I used is on page 38.

If you use different switches or buttons, or have a better layout in mind, go to town with it. Just draw a sketch first and make sure you include the measurements. Lay out everything carefully before you drill so you don't have to go back and fix anything later when you realize something doesn't quite fit.

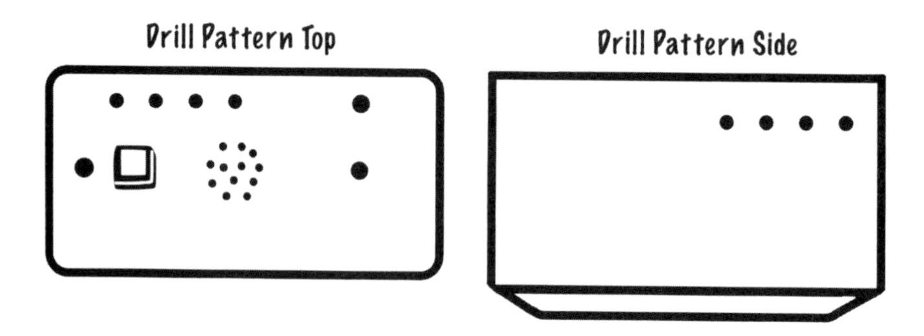

STEP 2: Mount the components you have that are not part of the cards into the box lid and side, according to your design. This includes the record selector toggle switches, ON/OFF button, the PLAY buttons, and the RECORD button.

STEP 3: Mount the power switch and the LED into the box.

STEP 4: Solder one end of the 100 Ohm resistor to the long leg of the LED (the Anode). Please note that this assumes you have purchased bare LEDs and need the resistor so it can handle the voltages we're passing through it from our AAA battery pack. It is also possible to purchase LEDs with the resistors already attached (see the ray gun project elsewhere in this book), which can save you this step.

STEP 5: Solder the other end of the resistor to one of the contacts on the switch. Solder a wire from this resistor switch junction and leave the other stripped end of the wire loose. This will be how we distribute power to the boards.

STEP 6: Connect the red wire from the AAA battery pack to the open contact on the power switch (the one without the resistor).

STEP 7: Connect a wire (I used black) from the short leg of the LED (the cathode) to the black wire on the battery pack. At this point, you should load up the batteries and hit the power switch. The LED should light up. If it doesn't, check your wiring and make sure the LED is wired correctly.

PERFORMING SURGERY ON THE GREETING CARDS

What comes next is very much like transplant surgery. We are going to have our new component box on one side, and a greeting card on the other. We'll open the greeting card, exposing the components

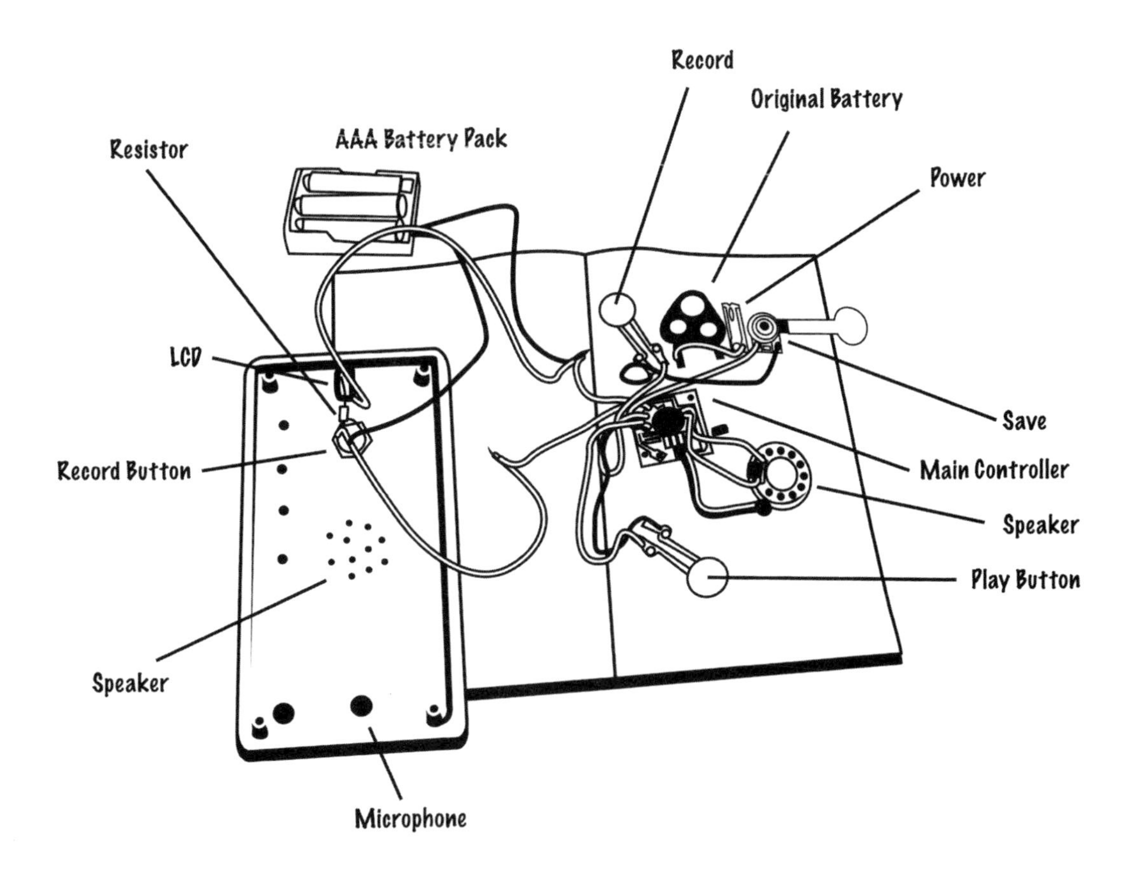

inside. Then we'll carefully start detaching components one wire at a time, reconnecting each one to a new switch, button, or power lead in the box until we've transplanted everything we need to make a new living, breathing monster. Or recording device. Whichever.

STEP 1: Expose the components in the first card without removing them. This is usually a matter of splitting open the card, perhaps with a hobby knife. (That's the generic name for an X-Acto knife.) Be careful not to break any wires, and go slowly to make sure when you pull things apart, glue doesn't cause any existing connections to be broken.

STEP 2: Take careful note of the different switches; making a diagram helps here. Better still, label them directly on the card and take a picture of it for reference. Do not remove the module from the card yet. It will be easier to liberate it in steps and remove it when all the new connections are made.

The cards we're using here have eight main components:

1. **Power Tab**—This is usually a plastic tab and very small circuit board; you remove the tab to activate the card. We will replace the four (one from each card) with a single ON/OFF button for all four controllers.

2. **Save Tab**—Also a plastic tab and board combo; you remove the tab to save the recording. We will replace these with switches used to indicate which controller module (see #8 below) the recording is being saved to.

3. **A Record Button**—We will replace these with a single RECORD button for all the modules.

4. **A Play Button**—We will replace each of these with a new button.

5. **A Microphone**—We will reuse only one of these (there are four total, one in each card).

6. **A Speaker**—We will reuse one of these (again, there are four total, but we will reuse only one).

7. **A Battery Pack**—This will be replaced with 3 AA batteries to power all the modules.

8. **The Controller Module**—A small board with chips on it that serves as the brain of the whole device (all the wires on the card feed in and out of this).

STEP 3: Detach the black wire from the small battery pack in the card and connect it to the black wire on the AAA battery pack.

STEP 4: There are two wires coming off the power tab. One goes to the board and one goes to the card's battery pack. Unsolder the wire that goes to the board from the power tab and connect that end to the black power distribution wire. With this step, the power switch and the batteries are now disconnected from the module.

STEP 5: The save switch is a simple tab that, when pulled, will disable the RECORD button. We want to be able to record over and over but still be able to disable recording. To accomplish this, we are going to replace the one-time-only tab with a toggle switch.

The SPST toggle switch (which you should already have mounted in your box) has two terminals. Detach one of the wires from the card's save tab and attach it to one terminal on the switch. Then detach the second wire from the save tab and connect to the other terminal on the switch.

You should be able to record a test message. If the message does not record, flip the switch and try again. If it still does not record, check your wiring.

The save tab is now disconnected from the module.

STEP 6: Now replace the PLAY switch with one of the push-button switches. This is a pretty straightforward task. The push-button switch has two terminals. Unsolder the wires from the PLAY switch, and attach each wire to one of the terminals. The PLAY switch may have some plastic wrapped around the terminals; if so, just cut it away with a hobby knife.

Make a test run of the play function. Make sure you are not touching the terminals when you test; the resistance of your skin could trigger the module.

STEP 7: Remove the wires from the RECORD switch and wire it to one of the momentary switches just as in the previous step. This will be a universal RECORD switch, so the rest of the modules' RECORD switches will be wired here. As always, test before you move to the next step.

STEP 8: Now remove the speaker from the card and mount it in the box. Depending on the strength of the attachment, you might need to use a hobby knife to cut away any glue. Once the speaker is free, unsolder the two wires from it (which were the old power leads).

Apply hot glue around the edge of the speaker and set it inside the box, over the grille holes you drilled. The speaker should not touch the box directly but sit on the hot glue.

Solder two new wires onto the speaker leads so that you will be able to connect all of the controller boards to the speaker. Then solder the original speaker wires to the new speaker wires. Hopefully, this isn't confusing. The idea here is that you disconnected the speaker from it's original controller board, added two new wires to it, and then reattached those wired to the controller board. The new wires will help to attach the speaker wires from each of the other controller boards as well, since you are sharing one speaker for all four boards.

Run a test. If the speaker makes a buzzing noise, that means it is touching the box, which is setting up a resonant buzz. We don't want this. Remount it and make sure it is not touching the box but sitting on a bead of hot glue.

STEP 9: Remove the microphone from the first board by unsoldering the wires from the board. Just as you did with the speaker, add a new wire to each of the mic's contacts, and then attach those wires to the controller board. Just as you did with the speaker, this mic will be connected for all four controllers to share.

Hot glue the microphone into the box in the spot you prepared for it.

You have now transplanted the first card. The remaining steps are to be carried out on the remaining three cards, which will each be sharing power, speaker, and microphones.

STEP 10: Remove the black wire from the module battery pack and attach it to the black wire on the AAA battery pack. Remove the white wire from the power tab and connect it to the wire coming from the power switch you mounted earlier.

STEP 11: Repeat Step 10 for each of the save tabs and the PLAY switches as described above, connecting those leads to the separate buttons you mounted in your box earlier.

STEP 12: Remove the RECORD switch from the card and disconnect the wires. Attach the wires to the RECORD switch you added earlier, and discard the old switch. There will be a wire coming from the controller board and one coming from the save switch. When you make the connections to the RECORD switch, make sure the wire coming from the new controller module is connected to the same terminal as the wire from the first controller module, so the positive and negative are aligned.

STEP 13: Disconnect the wires leading to the speaker in this card, and connect them to the speaker you installed in your box.

STEP 14: Disconnect the wires leading to the microphone and connect the new microphone wires to the microphone installed in the box earlier.

And with that, you should be done. Test, verify, and troubleshoot at every step so you don't leave a boo-boo along the way and not realize it until much later.

FINISHING UP AND PLAYING WITH YOUR NEW CREATION!

Your control setup is thus: There are four PLAY switches, four RECORD select switches (to determine which controller module it is being recorded to), and one RECORD button. To record, flip the RECORD select switch on the channel you want to use, and press the RECORD button.

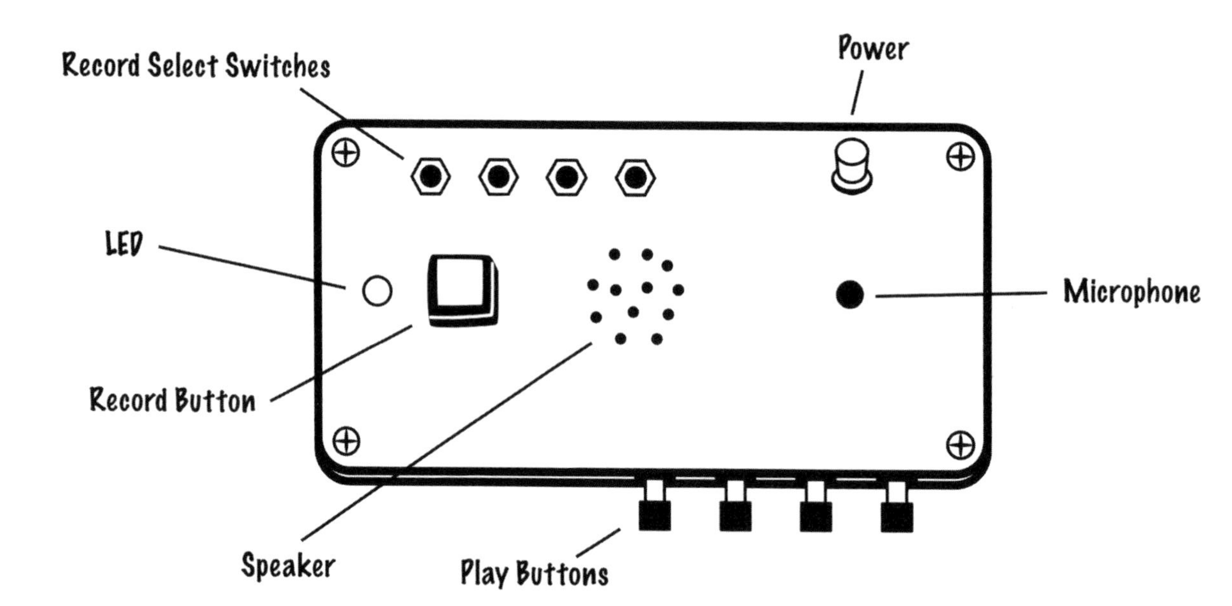

Hit the PLAY button for the specific module to play back what was recorded on it. When you have it just right, flip the RECORD select switch back to protect the recording from being overwritten. You can record multiple modules by flipping multiple select switches to active at the same time.

Finally, wrap any exposed connections in electrical tape, stuff all the guts into the box, and close it up. You have successfully created your own four-channel sound box, good for hours of silly fun!

Go Medieval at Home with Make-Your-Own Weapons

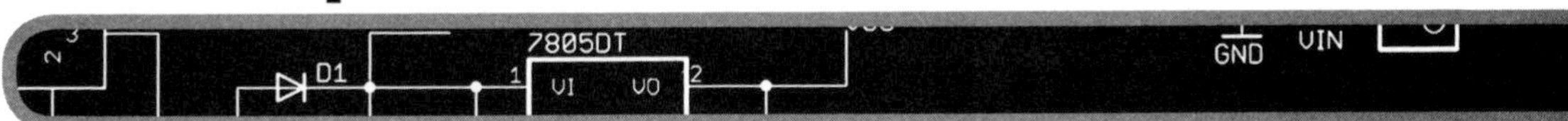

Celebrity Geek Dad Project by Chris Anderson

Chris Anderson is the editor in chief of Wired *magazine, one of the few print publications pulling itself successfully into the modern media era. He has written two* New York Times *bestselling books on business trends in the Internet era (*The Long Tail *and* Free*). He also founded the blog I now run (and which led to this book), www .geekdad.com.*

And he's the geeky father of five kids.

As parents, we try so hard to keep our children from fighting with one another. We do our best to instill in them the idea that when they get older, while friends and romances may come and go, siblings will always be there. And they do their best to deliver noogies and wedgies to one another.

In a household with five kids, the family dynamics that Chris has had to deal with can only be imagined. And so it's not surprising that the project he offered to develop for this book involves creating padded, "safe" versions of medieval weapons that kids can use to battle one another.

PROJECT	GO MEDIEVAL AT HOME WITH MAKE-YOUR-OWN WEAPONS
CONCEPT	Build padded versions of ancient weapons for kids to play-fight with.
COST	$–$$
DIFFICULTY	⚙ – ⚙⚙
DURATION	⏱⏱ – ⏱⏱⏱
REUSABILITY	🌐🌐🌐🌐
TOOLS & MATERIALS	½-inch PVC pipe; ½-inch (inside diameter) foam pipe insulation; cardboard; spray adhesive; Gorilla Glue or equal; duct tape; hacksaw or other; cardboard knife; sandpaper; scissors; nylon straps; plastic buckles

There are few things more guaranteed to be a kid pleaser than creating ways to hit one another. Why not harness this force and turn it into a great rainy-day craft project? Your kids can make foam swords, axes, and shields in less than an hour and be walloping one another (safely) after lunch. Best of all, these weapons are tough and will last years, creating a foam armory that can work for everything from pickup battles to outdoor live-action role playing to imaginative Halloween costumes.

All you need is a trip to the hardware store to pick up some foam pipe insulation and PVC pipe. Go for the ½-inch PVC pipe and tubular pipe insulation foam, which has an external diameter of about 2 inches—any thinner and it hurts, and any thicker and it doesn't really look like a sword anymore. You can also use a ½-inch wooden dowel in place of the PVC, but it's a bit heavier than the PVC pipe and can hurt a bit more as a result (it can also break if you hit something hard enough). They usually come in 10-foot lengths, and two of each will make at least four swords. If you want to make a shield, you'll need about 5 feet of foam insulation, so remember to account for that, too.

THE BROADSWORD

STEP 1: Swords should be between 3 and 4 feet long, depending on the size of your kid. Along with the desired blade length, you'll need about 10 inches for the handle, with room for the hand plus a couple inches for the butt of the sword and the hand guard on either side. Saw off the PVC pipe, at the length you've chosen, with a standard wood saw or hacksaw, and clean up any burrs or rough edges. (You can also use shorter pieces of pipe to make daggers.)

STEP 2: Using a hacksaw or a big pair of scissors, cut the foam insulation into three pieces: blade, butt, and guard. You can measure the blade part of the foam by laying it next to the PVC pipe. The foam for the blade should extend a good 2 inches farther than the pipe on the "tip" end (so nobody gets hurt if there's some jabbing going on), and leave 8 inches or so of exposed PVC pipe at the base. Then cut a 2-inch bit of foam for the butt of the sword, and another 5-inch piece that you'll use for the hand guard. Cut a ¼-inch hole through the side of the 5-inch piece, all the way through to the other side; you're going to push this onto the pipe crosswise to make the hand guard.

STEP 3: The best kind of adhesive is Gorilla Glue (because it foams up and expands into the gaps between the pipe and insulation foam). Squirt a pretty good glop down the inside of the foam tubing, starting at what will be the handle end (don't put it on the PVC pipe, because it won't spread evenly) and going up to within 3 inches of what will be the tip end. The tubular foam insulation is slit all the way up its length, so it's not difficult to access its interior.

STEP 4: With the glue in place, you are ready to put the PVC inside the foam. Lay your piece of PVC next to the foam so that the handle end of the pipe extends beyond the end of the foam by the aforementioned 10 inches, and the tip end of the PVC comes short of the tip end of the foam by 3 inches. Push the PVC pipe into the foam through the slit, starting with the handle side of the sword, until it is completely enveloped by the foam.

STEP 5: Apply a bit of Gorilla Glue to the PVC exposed just below the foam, and push the hand guard piece of foam onto the pipe at the butt of the PVC through the side hole you've cut in the foam (the foam will be going on sideways, perpendicular to the blade). Push it up to where the pipe is glued, and tight against the base of the foam blade.

STEP 6: For the grip of the sword, you can wrap the PVC pipe with duct tape. Then glue on the last 2-inch bit of foam at the end of the pipe to form the butt of the sword. To neaten things up, cut a little more foam and glue it in the 2-inch hole at the tip of the sword, and stuff a little in the open end of the PVC at the butt.

Optional Geekiness

You can wrap the entire sword in silver or other color duct tape to give it an awesome look and an even more solid sound and feel. You can also, prior to putting the foam on, put dry rice, beans, sand, or even bells into the PVC shaft and stop up the ends so that when you wield the sword in combat, it makes mystical sounds.

THE SHIELD

STEP 1: For the shield, you'll need two pieces of corrugated cardboard cut to the size you want your shield to be (an 18-inch-diameter circle is a good start). Cut out the two circles, then spray one side of each with 3M spray adhesive (if you can't find spray adhesive, a layer of Elmer's Glue on each side will work; just let it get a bit tacky prior to laminating). The important thing when you're gluing the two sides of the shield together is that you rotate the cardboard so that the "grain" of each side (the direction the corrugations go) are 90 degrees apart, so one goes one way, and the other is perpendicular to it. This will strengthen the shield so that it won't have the tendency to fold along the grain of cardboard in any one direction. This also is why circular shields are easiest to do: The two pieces of cardboard will match each other even after one is rotated, without your having to think about it during the cutting process. Press the two sides together, and then put them under some books to dry flat, according to the directions on the packaging of the glue you used.

STEP 2: Once the shield is dry, you can make the handholds. The best things for this are nylon straps, which you can also get at the hardware store, along with backpack-style plastic click buckles. Have your warrior place her or his hand on the back of the shield where it will be held, and mark two sets of lines, one pair on either side of the palm, and the other pair at the wrist. Then, at each of those lines, cut a slit the width of your nylon tape through the cardboard. Thread the nylon strap through and attach the buckles, leaving enough extra strap so you can adjust the size.

STEP 3: Finally, it's time for the padded rim. Measure a length of foam insulation tubing to match the circumference of the shield. Split the foam down the seam, and press the edge of the shield into this split at one end of the foam tubing. Use short strips of duct tape to adhere foam to cardboard, but don't cover too much of the front (since you're going to want to decorate it). Now work around the rim of the shield, pushing the foam onto the shield, and duct-taping every 6 inches or so. When you get to the end, you'll find that you came out a little

long on foam. Trim it to fit and duct-tape the join. You now have a foam rim around the entire shield.

STEP 4: Now it's time to decorate the front with markers and more duct tape. Be creative! This might be a good time to figure out what your family coat of arms is going to be. A little Web search will provide good ideas, and you may find that your family name already has a coat of arms, which you can use as inspiration.

Extra Geeky Ideas:

There is no end to the other foam weapons you can make. Some sofa foam (also called open-cell foam, available at your fabric/crafts store), carved into shapes with an electric carving knife, will make a great axe head (and use a padded PVC shaft made just like the sword above so nobody gets hurt). You can make foam-tipped spears and arrows, and even a cool two-sided lightsaber, minus the lights.

If your kids want to go all the way down the rabbit hole on this, there's always live-action role playing (LARPing), which adds ever more elaborate costumes and themes, including armor, along with equally elaborate rules. It's incredibly nerdy, but hey, they're outside!

Smartphone Steadicam

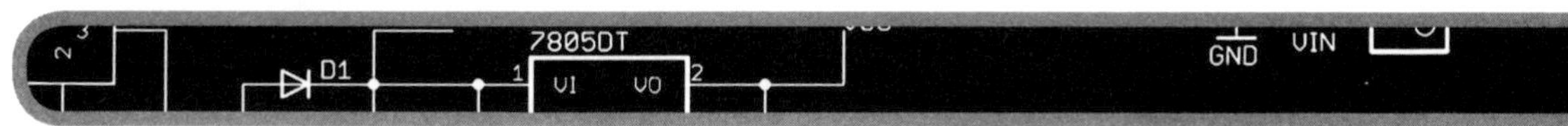

We live in the future. We may not have jet packs or transporters yet, but we do carry around in our pockets computers more powerful than the onboard systems on Apollo 11. And by "we" I mean you, the geek parent, and very likely your kids as well. More than a computer, your phone also acts as, well, a phone (duh), a media player, and even a camera. It's the last function that's of interest to us in this project.

We've heard dark tales about overbearing governments that used ubiquitous video cameras to keep tabs on their cowed population. But these days, something revolutionary has happened since video cameras are small and cheap: Everyone can carry them, and anyone can record events anywhere, anytime. It's a powerfully equalizing force.

It's also a powerfully inspiring creative force.

Having that video camera everywhere you and your kids go is handy for capturing the important moments of your life, but one of the drawbacks of the camera phone compared to larger, dedicated camcorders is that you can't mount them on a tripod for stability. Or, beyond a tripod, you also can't use a Steadicam rig. There's no way to avoid that jumpy, inevitably amateur-looking video.

"What's a Steadicam?" I hear you ask! Obviously we need to work on your film-geek credentials. For professional filming, a Steadicam rig is a fancy, hydraulically augmented frame worn by a camera operator who can then walk, or even run/hop/skip/jump around, without the movement jittering the camera. It's a powerful tool used for smoothly capturing action shots.

Even cooler is that many people have developed DIY Steadicam rigs for home video cameras. For less than $50 for materials plus a little work, you can build your own rig that helps get that smooth, Steadicam look. These rigs usually use PVC or metal piping and counterweights to counteract the camera user's movements as they record something.

It occurred to me that with everyone using smartphone video cameras to record snippets of life, there needs to be a Steadicam rig for smartphones to improve the footage taken with them. And so, with some easy-to-find and inexpensive parts from the hardware store, I came up with a fun project to do to make that exact device!

PROJECT	SMARTPHONE STEADICAM
CONCEPT	Build a rig for your smartphone to help smooth out action-shot videos.
COST	$–$$
DIFFICULTY	⚙ — ⚙⚙
DURATION	☼☼ — ☼☼☼
REUSABILITY	🌐🌐🌐🌐
TOOLS & MATERIALS	½-inch-by-16-inch threaded rod; ½-inch-by-8-inch threaded rod; small clamps and other, larger clamps to secure the build while glue dries; ½-inch regular nuts; 20 ½-inch stamped washers; ¼-inch stamped washer; one bottle strong epoxy or superglue

We are making something like a monopod, which is like a tripod for a camera but with one leg instead of three. It will have clamps on top to hold the smartphone, since a smartphone doesn't have the little screw hole for a standard camera mount, and it will also have a counterweight at the other end to help buffer the jittery movements that usually transfer from your legs through your arms and hands to the camera when you use it naked (the camera, not you). We'll add a perpendicular arm off the side of the rig to help you control rotation of the camera as well. Hopefully, when you put this contraption together, the video you'll record will look quite a bit smoother (and help make your child's career as a YouTube producer a success!).

One warning about this project: There's a fair bit of gluing and waiting for glue to dry. So it may be wise to tackle this project in tandem with another one so that, as you achieve a step on this project, you can set it aside while you go off and enjoy another fine project from somewhere in this book.

STEP 1: Gluing the Clamps

What we're doing first is affixing the clamps to the nuts that will eventually mount onto the threaded rod. Because you're gluing them, they're going to have to set for some time to allow the glue to dry.

How you put these together will depend on the exact clamps you get, but I'll describe this project based on using locking caliper-style clamps. In this scenario, because each clamp has two arms, use two nuts set parallel, one on each arm. Apply the glue to each surface per the directions on the packaging, and make a nut sandwich with the clamps as the bread. Then use additional clamps to hold the construct together so it can dry.

Because of the necessary drying time, let's glue one more thing. Take one nut, lay it on edge on your work surface, and use your glue to affix the ¼-inch washer flat to the top edge. Then take another one of the nuts and glue it flat onto the top of the washer (so the whole is pointing up, turned 90 degrees from the nut below it).

Follow the instructions on your glue packaging and allow everything to dry well.

STEP 2: Assembling the Counterweight

Choose a "bottom" and "top" end of the 16-inch threaded rod. Take one of the nuts and run it up the rod about 4 inches from the bottom.

Now take all 20 of the ½-inch washers and fit them onto the bottom of the rod up to the nut. Finally, take one more nut and run it up tight against the washers. There should not be much of the threaded rod sticking out below the bottom nut. You can put a dab of glue where the nuts and threaded rod meet so they'll be permanently stuck, but that's not absolutely necessary, especially if you use a couple wrenches to twist them tight instead.

What you've just done is to build the counterweight for your Steadicam. The counterweight should be as heavy, or even heavier, than the camera you'll mount on top, since it's going to help absorb shakes and also let you record more smoothly if you need to move the camera.

STEP 3: Assembling the Handle

When the nut-washer-nut sandwich in Step 1 is dry, you can run it down the threaded rod from the top, to about one-third of the way from the top. Hopefully, it's obvious which way it goes on—only one of the two nuts will allow you to go that far down the threaded rod.

Now take the shorter threaded rod and insert it in the nut that's facing out 90 degrees from the sandwich you just attached. This is your handle. You can take some additional nuts and run them down the outer end of the handle so you have a better surface to hold than the threaded rod, but this is a personal choice.

Once again, you can apply glue to each of the nuts that are loosely screwed onto threaded rods and let them sit so everything will be firmly attached. However, you can skip this for the sake of time, or if you want to be able to break down the whole assembly and carry it, in a more portable fashion.

STEP 4: Finishing Touches

All that's left is to take the clamp assembly and screw it onto the top of the threaded rod. It doesn't really matter which of the two nuts you use. Just run them down until the top of the rod is flush with the top of the nut.

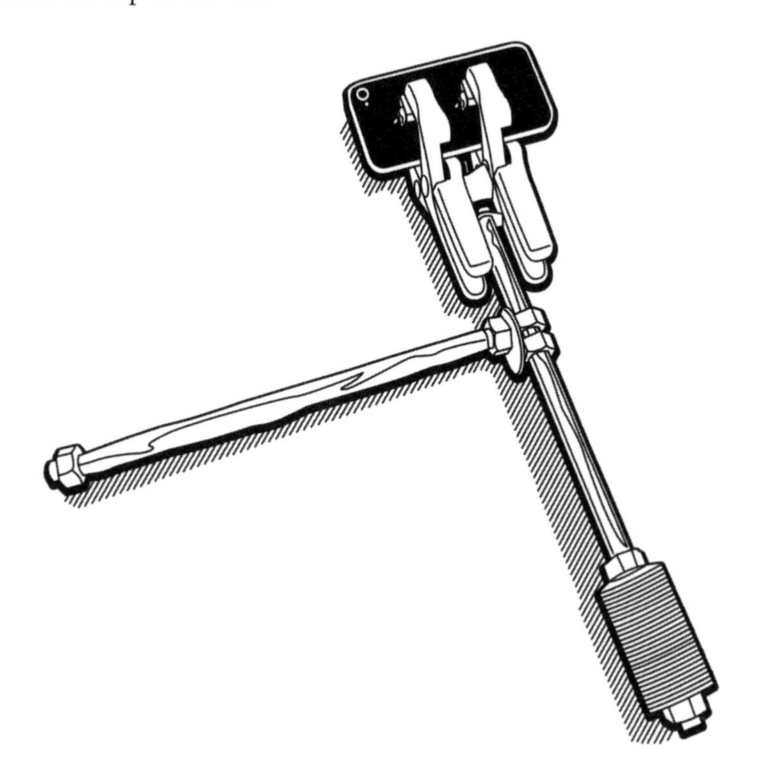

Take your camera phone, set it to video mode, and put it into the clamps, closing them tightly enough to make sure the phone won't shift, but not so tightly that you feel like your going to break the glass or case. Depending on the model of your phone, you may need to make sure you have something rubbery to keep the phone from sliding. I tested this with an iPhone 4 and it did tend to slip, so I picked up one of those rubber circles people use to open jars, cut out some smaller circles and glued them to the feet of the clamp. It did the trick perfectly!

To use your new device, hold the Steadicam rig with your primary hand gripping the nuts where the handle tees into the vertical threaded rod. Place your other hand at the end of the handle. Keep both arms extended, and slightly bent at the elbow, and also walk with your knees slightly bent so that every joint of your body absorbs the jostling of walking. In no time, you and your junior cameraman both will master smooth-action video!

Skitterbot!

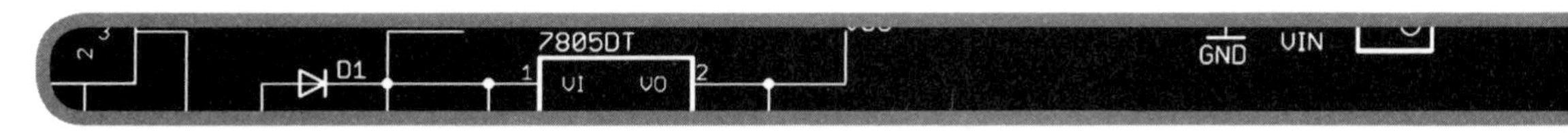

(Project Idea by Anton Olsen)

Robots fascinate us geeks, and have from the very beginning. That is, since 1920, when Czech playwright Karel Čapek first used the word in his play *R.U.R.* (aka *Rossum's Universal Robots*). Karel's brother, Josef, coined the word for him, which was derived from the Czech word for "work," or "serf labor." Thus, the original idea of a robot was a mechanical creature that performed menial tasks on behalf of mankind.

But we, being the pessimistic humans we are, always have to push good ideas to their worst possible conclusion. And so robots are not always portrayed as creatures created with the most positive purposes in mind (the *Terminator* movies are a prime example), or their purposes are positive for a greater good that doesn't just include humanity (see *Silent Running*). The typical story line is that they first exist to end the backbreaking labors humans must endure, to free us for more lofty pursuits, but then everything goes wrong, the robots take over, and we end up becoming the servants. Or worse, the fuel. Or even worse: We become completely unnecessary.

But until the robot Skynet from *The Terminator* becomes self-aware and takes over the planet in real life, we have some time to enjoy what robots can do for us. And in the spirit of hacking and building, here is a project for you and your kids to build a simple robot for your own enjoyment. At least until Kyle Reese shows up.

PROJECT	SKITTERBOT!
CONCEPT	Build your own remote-control robot from some scrounged parts and materials.
COST	$$–$$$$
DIFFICULTY	⚙⚙ – ⚙⚙⚙
DURATION	☼☼☼ – ☼☼☼☼
REUSABILITY	🌐 – 🌐🌐🌐🌐
TOOLS & MATERIALS	**Materials**: RC transmitter; matching miniature RC receiver; battery for the receiver; 2 cheap hobby servos; foam core or heavy cardboard; tape, double-sided foam tape or rubber bands; paper clips **Optional materials**: hot-glue gun; markers; googly eyes; pipe cleaners; self-sticking rubber feet **Tools**: pliers or wire cutters; small nail scissors; hobby knife

Now I'll come clean: In proper parlance, this is not technically a robot. It's really just an RC toy. What's the difference? Well, you can't program this device (at least not in this design), so it can't work autonomously. But that's just being a word nerd, and our point here is to have fun. So if we want to call it a robot (or a Skitterbot), we'll darn well do it!

The Skitterbot itself is an interesting machine—sort of a buglike creature that drags or pushes itself along on the floor, using the motion of servos in two planes to do the work. What are servos? They're sort of like small motors, but instead of causing wheels or gears to spin around, we'll use them to make foam-board arms move back and forth to move the bot.

You can think of Skitterbot sort of like a sled with arms that reach forward and grab the ground to pull the body along. However, in addition, there is powered articulation between the main body and the back runners of the sled so that between the pulling/pushing arms and the wiggling of the body, a more dynamic range of motion is possible.

PICKING THE RIGHT PARTS

The cost of this project can vary widely, depending on what you have available already in your home or can buy cheap. For the radio controller, I recommend looking for an inexpensive two-channel 75 Mhz aviation bundle with a transmitter and receiver, available in many hobby stores. Aviation transmitters with dual joysticks will be easier to control, and their receivers are usually smaller and can run off smaller batteries. Cost is as low as $40 or as much as $100. The servos can be found for $10 each or less. If you already have the remote at home, then you may be able to save some money—just make sure you get the correct frequencies so that all the parts will work together.

Just What Is a Servo, You Ask?

A servo is a small electrical motor that turns a shaft—sort of like the drive train in your car. What's important for the purposes of this project is that, unlike a regular motor, the servo does not spin the shaft. Rather, it turns it through a limited rotation in one direction, and back to a neutral point, or in the opposite direction and back to the neutral point, depending on how the electrical current is applied (positive will go one way, negative the other).

The size of the servos will determine how big and heavy your robot can be. Smaller servos may be cheaper but may move less weight. You'll need to do some tests to find a balance between overall body weight and the strength (and therefore cost) of your servos. But if you're lucky, you won't have to compromise too much. Most servos will be able to handle the basic build we're doing here. Just keep this in mind if you plan to . . . "augment" the design of this robot.

GETTING THE PARTS READY

Assemble the servos with the X-shaped arms. We want one pair of arms to be perpendicular to the body of the servo. You can center the servo shaft by plugging it into the receiver and applying power and then releasing to get it to its "neutral" position. Add the X-shaped arms so that one of the lines of the X is in line with the body of the servo. This will ensure that when you move this servo's remote control stick in one direction, it'll turn 45 degrees one way. Then when you push the joystick all the way in the other direction, the servo will turn back through the neutral position, and 45 degrees the other way. This will create our push-pull action.

The other thing to think about before designing the body and assembling the robot is the power source. You'll probably be using the rechargeable battery pack that came with the transmitter and servos. You should plug this in and make sure it's charged up so you'll be ready to go when everything is set.

BODY DESIGN AND CONSTRUCTION

Start by laying out the servos, battery, and receiver on the piece of foam core or cardboard to conceptualize the spacing. You won't need too big a piece, maybe 4 inches by 8 inches, and if you are using a small battery, the battery and receiver can stack on top of each other, so they will fit in the same horizontal space.

The forward servo will be mounted with the shaft perpendicular (pointing down) to the ground. This will drive the front legs that pull the robot along as it crawls. The rear servo will be mounted with the shaft parallel to the ground and is used to shift the robot's weight from one side to the other. Somewhere in the middle will be the battery and receiver.

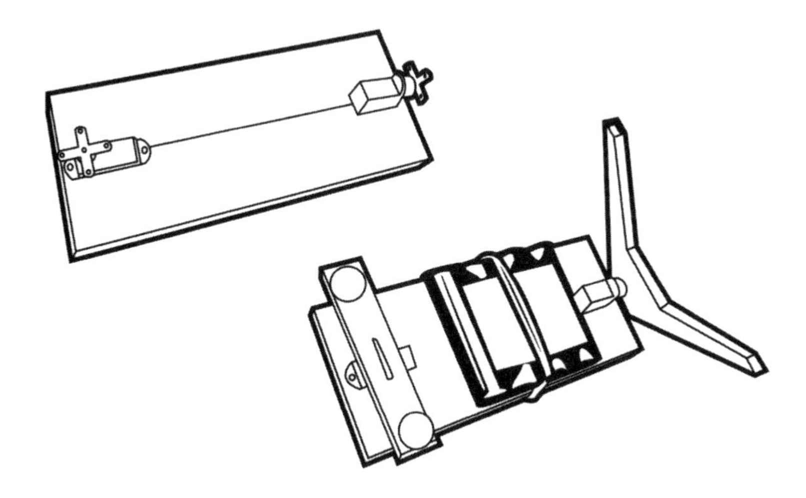

This is roughly how long the robot body needs to be. In this basic design, I show a simple rectangular body about 4 inches by 8 inches, but you can make it any shape large enough to hold all the components. Just remember, it is always easier to make it shorter, so err on the long side. Oh, and "measure twice, cut once!"

Hold the base of the forward servo near the front center of the body, with the shaft straight up in the air, and carefully trace around the servo. With a hobby knife, cut out a rectangular hole slightly smaller than the servo. If the fit is snug enough, you'll be able to just slip the servo into the hole you made and it'll be tight enough to stay in place so you don't need any tape or glue. Otherwise . . . tape. Or glue.

Mount the rear servo with the shaft parallel to the ground and pointing to the rear of the robot. Align the servo so the shaft extends past the rear of the body. Trace around the body of the servo, paying close attention to the fact that the servo's body in this aspect has tabs extending outward. Cut a slit in the foam core to accommodate the tabs, and cut out the slightly smaller space for the rest of the servo body to wedge into (again, hopefully, tight enough so that you don't need tape or glue). Press the servos into place.

In this basic design, attaching the receiver and battery is almost comically simple. Place them in the middle of the body (as determined by your original layout) and affix them with two-sided foam tape, or rubber bands, or both (for extra stability). There, done!

DESIGN AND CONSTRUCTION OF THE LEGS

To make the simplest front legs, cut out a bar of foam core measuring about 1 inch by 4 inches. Then add a couple rubber bands on the ends for traction. Since these legs are easy to make and attach, you can experiment with different shapes to see which looks and works the best. A great alternative to rubber bands for the "feet" are the little peel-and-stick rubber feet you can get for the bottoms of vases to protect shelves and tabletops.

The rear legs should be a little taller (say 3 inches), a little wider (try 5 inches), and should look like a very wide inverted V or chevron (the insignia, not the oil company). Other shapes work, but the rear legs will be dragging on the ground so achieving less friction by making the feet very small (like the tips of the ends of an inverted V) is better.

The legs are attached to the servos using a bit of bent paper clip or wire. The servo arms each have a little hole in them. Quite conveniently (almost as if they were designed that way!), this will allow us to connect them to the legs.

Make a square U out of a paper clip, with the vertical sides the same distance apart as the widest holes on your servo arms. Poke the two ends of the wire through the foam legs and the holes in the servo arms, then bend outward to hold them in place.

The front legs (the 1x4-inch bar with the rubber feet) mount to the front servo arms so that they are parallel and square to the front of the body. At rest, the weight of the front of the bot will sit on the feet.

The back legs mount to the servo arms so that the tips of the V point down (why we called it "inverted"). Keep the butt end of the robot raised off the ground. Each of the tips will act as a slide, so having as little surface as possible for these means as little friction as possible.

WIRING THE ROBOT

You are going to connect the servos so that they are controlled by separate axes on the same joystick. For some airplane receivers, this will be the elevator (the horizontal flaps on a plane that point you up or down) and aileron (the vertical flap that turns you left or right) channels. You may have to experiment some to find the correct ones based on your equipment, but, simply speaking, there will be two wires from the receiver representing "up" and "down" that will connect to each of the two contacts on one servo, and two wires

representing "left" and "right" that will connect to two contacts on the other servo.

Once wired, your robot should be placed on a smooth flat surface (hardwood floor works great), power everything on, and move the joystick around in circles. If everything works as it should, your robot will walk in one direction. Reverse the direction of your circle, and the robot will walk the other way. Try different patterns to make the robot turn or dance.

CUSTOMIZING YOUR SKITTERBOT

You can have some fun with this basic design and dress it up a little however you wish. Use markers to color the body and legs. Add some googly eyes and pipe cleaners for flair. Experiment with different shapes for the legs and body. Longer front legs will move faster; shorter ones can carry a heavier load.

Consider the issue of friction on the back end as well. What if, instead of the V-shaped legs, you could mount an axle with wheels? This could make the choice of surface less of an impediment.

There's plenty of room for your own personal modifications and adaptations with this design, so once you've built the base model, you'll certainly have some replay possibilities. So get building, and post your designs on www.geekdadbook.com!

Make Your Own Ray Gun

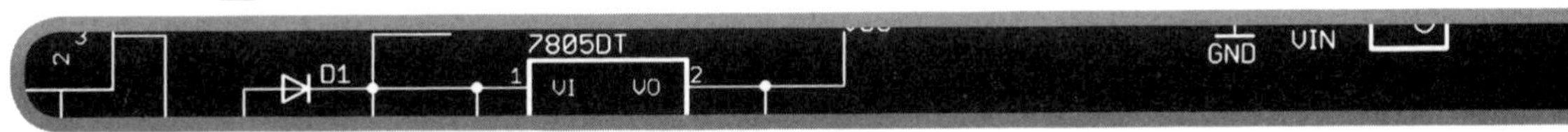

Whether you want your phasers set to stun, or you fully subscribe to the idea that "religions and ancient weapons are no match for a good blaster at your side," I think we can all agree that geeks who love science fiction love their energy weapons. From the pulse rifles in *Aliens* to the particle beam cannon in *Babylon 5*, future-tech weapons that shoot energy of one kind or another are way more exciting than boring old projectiles that shoot metal.

Of course, because movies and television must have cool visual effects, sometimes the science of these kinds of weapons gets twisted into fiction. For example, since the output of any energy weapon, from a basic laser to more complex energy weapons, should travel at the speed of light, what we see on the screen isn't the way these weapons would really work. For any distance less than a few miles, the time it would take to lance through a target after pressing a trigger would be nearly instantaneous. Characters wouldn't be able to react quickly enough to dodge the laser. Jedi couldn't raise a hand and deflect blaster fire, and Stargate team members wouldn't be able to jump away from zat fire. It just wouldn't happen. But then, of course, without these special effects and action scenes, the shows wouldn't last very long.

But even though we can't make real weapons that travel at the speed of light, we can have a lot of fun using our imaginations and making our own. Why buy when we can build?!

PROJECT	MAKE YOUR OWN RAY GUN
CONCEPT	Use some inexpensive electronic components and materials you can find around your house to build your own ray gun, complete with lights and sounds.
COST	$$–$$$
DIFFICULTY	⚙⚙ – ⚙⚙⚙
DURATION	☼☼ – ☼☼☼
REUSABILITY	🌐🌐🌐🌐
TOOLS & MATERIALS	9 volt battery w/ module; sound module; button module; LEDs; wire; LEGO bricks

There are two big parts to this project: building the electronics and building the structure of your ray gun. The first part will use some pretty specific components, but when you build the structure of the gun, you'll be able to use a lot more creativity (which is the best part of this project, I think!).

BUILDING THE ELECTRONIC GUTS OF THE RAY GUN

What are the basic effects of a ray gun? Light and sound, both of which can be achieved pretty easily using over-the-counter electronics parts.

While I was doing research for the Make Your Own Mini-Me project, I discovered that the world of model trains has much to offer, in the way of simplified electronics, to help you light things up and make noise. They are gadgets most folks use when building elaborate miniature towns. What I found was that, instead of having to buy straight LEDs and attach them to resistors (to allow them to run off batteries), many model train Web sites will sell you LEDs with the resistors already attached. They also sell sound modules for making noises like police car sirens (and, thankfully, a phaserlike sound), and they offer battery modules that let you hook it all up together. The materials I used for my version of this project all came from www.modeltrainsoftware.com.

We're going to build a pretty simple circuit that starts with a 9 volt battery. From the positive electrode, we'll attach to the positive side of the sound module (which also has its own switch that we'll be able to use as a trigger). The negative lead from the sound module will tie into the positive leads for all the LEDs we use. You have a lot of choice here. First, you can choose the LED. It's good to think ahead a bit about how you envision your structure, since that may affect the color of the LED you want to use. In my build, I used translucent LEGO bricks for color, so bright white LEDs were good enough for me. But you may want to try a completely different structure for your ray gun, and having specially colored LEDs may add to the effect. What's also great is that because LEDs have extremely low power requirements, one 9 volt battery can power a large number of LEDs for quite a while. I ended up using six LEDs in my small ray gun, but you could run ten times that many if you want.

And from the negative leads of the LEDs, we'll go right back to the battery. Simple! It's like water flowing around an Escher-designed stream. However, easier said than done. Let me give you a few more tips you'll need.

For the battery, use a battery module not unlike what you'll find in some children's toys. It has a small black holder with cap and foot electrodes that the 9 volt battery can snap into. This seats the battery very nicely and has a couple wires coming from the electrodes to make connections simple.

Where wires connect to wires, you have to do a decent job of actually, you know, *connecting* them. It helps to strip off at least a half inch of sheathing so you can put the ends of the wires right next to each other (side by side, not tip to tip), and then twist the heck out of them until they are very intertwined. Next, cut yourself a little electrical tape and wrap it around a few times. This will help the wires stay connected, and insulate them from the other contacts.

With all your connections made, you ought to be able to test that your circuit is working by putting the battery in and pushing the button. If it's not working, check your connections. That's *always* the problem (unless it's a dead battery). While electronics can seem daunting, there really are some very simple principles behind them, and knowing those principles can help you troubleshoot.

Quick Science Lesson

Electrical current flows. Heck, that's why it's called *current*! It flows like water, and there are direct analogies between the equations for electricity and hydrodynamics. The amperage of current is like the speed of water in a channel. The voltage of current is like the cross-sectional area of the water in a channel (a sideways slice through the water). And power, which is amps multiplied by volts, is basically the same as the speed of the water passing through that cross section, which is the volume of water that's delivered in a given second. So imagine 15 amps of 120-volt electricity (what comes out of most power outlets in a home) being able to knock you over in the same way that a 10-inch-by-12-inch-by-1-foot-thick wall of water traveling at 15 miles per hour could.

Well, that last part doesn't really work for the analogy—how electricity hits you is quite different from how water does, but you get the idea. Current is current.

BUILDING THE BODY OF YOUR RAY GUN

Now that you've built the guts of your ray gun, you get to be creative. While I'm here to give you an idea of what to build, you're going to want your and your kids' ray guns to reflect your personalities—kind of like a Jedi builds a lightsaber as the last test to become a Master.

My method uses the most awesome building tools known to geeks—LEGO bricks. I did not go buy any special bricks for this. I simply built with what I had on hand, and I used only standard pieces from standard sets—no Mindstorms or other special pieces.

The only constraints you have in creating your own design are that you must build a box into which the battery module can fit, with a place for the button to stick out so you can press it easily, and with a location for the LEDs to protrude so they can create the visual effect.

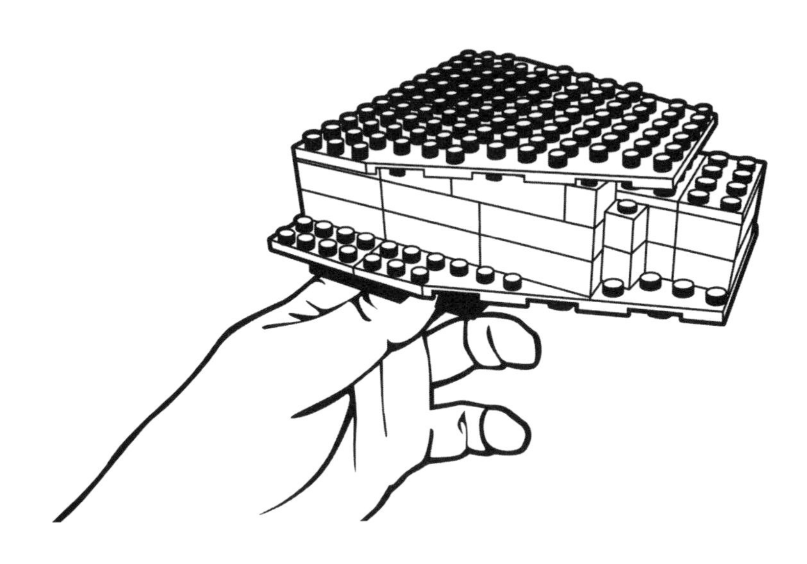

I was lucky enough to have some translucent bricks, so through them the LEDs add a nice reddish glow. My trigger button just pokes out the back, so it acts more like a thumb trigger (you'll notice I glued a small flat brick on top of the button to make the design consistent in look).

I'm sure folks can do a lot more, and be a lot more creative than this. If you're using LEGO bricks, try designing so you could actually have the trigger where the trigger goes. If you don't want to use LEGO bricks, well, that's not a problem, either. I can imagine all sorts of steampunk designs. You could even use the leftover copper pipe from the NERF Dart Blowgun project. Or something else entirely. It's all up to your imagination!

#47 06-22-2018 03:22PM
Item(s) checked out to p11362380x.

TITLE: A wrinkle in time [videorecording
BARCODE: 36644001916694
DUE DATE: 06-25-18

TITLE: Paw Patrol - Sea Patrol [DVD].
BARCODE: 36644001902298
DUE DATE: 06-29-18

TITLE: Geek mom : projects, tips, and ad
BARCODE: 36644001708125
DUE DATE: 07-13-18

TITLE: The geek dad's guide to weekend f
BARCODE: 36644001851354
DUE DATE: 07-13-18

Programmable Light Strings

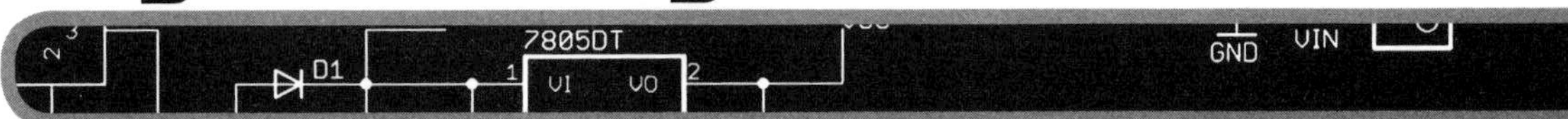

(Project Idea by Dave Giancaspro)

Back in the first Geek Dad book, I included a little project called Electronic Origami to introduce your children to simple circuits. I also included a project that used the very basics of the Arduino open-source circuit-board system to build a cool ambient lamp with repurposed CD/DVD discs. This project combines a little bit of both of those projects to build something fun and recognizable. But you'll be building and programming it with your own hands.

We've all bought inexpensive decorative light strings and hung them in our house for one reason or another. Obviously, these light strings get used a lot for holiday decorations, but sometimes these pretty lights are a great accent for a backyard party, or look nice hanging around the walls of a room as well. These days you can find really cool-looking light strings with special shades over the bulbs. My favorites are the ones that are made to look like hanging oriental box lamps. Those lights are the inspiration for this project.

PROJECT	PROGRAMMABLE LIGHT STRINGS
CONCEPT	Build a string of lights with origami lanterns over them, which you can program for any sequence you like with your computer.
COST	$$–$$$$
DIFFICULTY	⚙⚙⚙⚙
DURATION	⏱⏱⏱ — ⏱⏱⏱⏱
REUSABILITY	🌐🌐🌐🌐🌐
TOOLS & MATERIALS	7 feet of 8 Conductor Ethernet Network Cable (e.g., Cat 6); 12 LEDs (colors optional); 12 sheets of origami paper or semi-opaque colored paper; one Arduino Duemilanove; breakaway male headers (Sparkfun PRT-00116); soldering iron; solder; wire cutters/strippers

This is another one of those projects in which it would be far easier (and much less fun) to go out to your local big-box store, plunk down $15, and buy the premade product. But where's the geek cred in that? Furthermore, what you'll build in this project is fully customizable, from the shades on the LED lights to the sequencing of the blinking lights themselves. This project is the perfect example of the Geek Dad philosophy—it's important to truly know how a thing is made. By knowing, you're owning it!

Of course, in the process, you may also develop a newfound respect for automatic manufacturing processes and how they save us time and effort!

The general idea here is that you'll make your own string of lights by connecting individual LEDs to a power source and to the Arduino controller board, which you'll be able to program. This is going to take a lot of cutting and soldering of wire, which is why in general I suggest you do this project with older kids. However, if you watch them closely, it could also be a good project for teaching younger kids how to solder. Since the wire-to-wire connections don't have to be as tidy as those connected directly to a circuit board, you can work on technique and safety. (Two wires only have to touch and conduct electricity, so lots of solder doesn't really matter when connecting them; whereas, if connections on a circuit board have too much solder, you'll get overlapping connections and short circuits.)

About Arduino

As I mentioned, I introduced readers to the Arduino open-source programmable circuit boards in my first book. It is a very cool open-source initiative that puts practical programmable electronics into the hands of hobbyists. Check out www.arduino.cc for some great information about Arduino.

The boards can be built by hand or purchased preassembled, and the software can be downloaded for free. You can buy an Arduino board, hook it up to your home computer, download and run some free software, and actually program the chips on the board to do different things with modules you attach to the board. An Arduino board will help you teach your child that all the chips and wires crammed into every piece of home electronics you own aren't really magic boxes, but instead they are simple devices that you can easily learn how to hack with the right tools.

PREPARING THE STRING OF LIGHTS

STEP 1: Start out by cutting a 6-foot length of the network cabling. You can clip the jacks off because we will not be using them. Obviously, this project is designed to use a specific length and spacing of the lights. This is completely arbitrary, and you could adapt these directions and go for a longer length with greater spacing.

STEP 2: Strip the outer layer of insulation from the Ethernet cable and discard. This will leave you with four "twisted pairs" of cable (so eight overall lengths of wire).

STEP 3: Untwist one of the pairs and clip the white wire into 6-inch pieces. You should have 14 pieces when you're done.

STEP 4: Strip ¼ inch of the insulation off each end of each white wire, exposing bare wire. Divide the white wires into two groups, one with 11 wires and one with 3.

STEP 5: Take one of the 11 wires and solder one end of it to the cathode of the LEDs. The cathode will be the short leg of the LED, and is the negative electrode. Oddly enough, on circuit diagrams, it is noted with a *k*.

STEP 6: Solder the other end of the wire to the cathode of the second LED.

STEP 7: Take the next wire and solder one end to the second LED cathode (you should have two wires there now) and the other end to the cathode of the third LED. Continue doing this until all 12 LED cathodes are connected together in a string.

STEP 8: Next you will need to cut 12 pieces of solid colored wire (i.e., not white) with the following measurements from the rest of the Ethernet wire (specific color for each piece does not matter—just keep track of them by size):

1 at 6 inches, 1 at 12 inches, 1 at 18 inches, 1 at 24 inches, 1 at 30 inches, 1 at 36 inches, 1 at 42 inches, 1 at 48 inches, 1 at 54 inches, 1 at 60 inches, 1 at 66 inches, and 1 at 72 inches.

Strip each end of each piece of wire so ¼ inch of bare wire is exposed.

PREPARING THE CONTROLLER BOARD

STEP 1: The breakaway male headers are a standard component for working with circuit boards. They insert into the header modules on the Arduino board, so we'll have contacts to which we can solder our wires, allowing the current controlled by the board to flow—as programmed—to the lights.

The part specified in the list of materials is a strip of 40 male headers that will snap apart into smaller lengths, depending on the circuits being worked with. For our purposes, break off one length of 7 pins (meaning one contiguous bar with 7 contacts) and one length of 8 pins.

STEP 2: On the Arduino board, there are two header modules, each with eight receiving holes. Each of these holes has a corresponding designation printed on the board itself. The first module should have holes 0 through 7, and the second module has holes 8 through 13, plus GND (for "ground") and AREF (for "Analog Reference"). Put the 8-pin male header, which you just made, into the module with holes 0 through 7, making the long pins go down into the holes. Put the 7-pin male header into the other module, filling holes 8 thorugh 13 and GND. We will not use the AREF hole for this project.

HOOKING EVERYTHING UP

STEP 1: Take one of your three remaining short white wires and solder one end to the Arduino GND pin.

STEP 2: Now it's time to attach your twelve variously lengthed wires to the contacts on the Arduino board. You're going to solder one end of each wire to a specific contact for one of the holes on the header modules:

Wire Length	Header Position
6 inches	0
12 inches	1
18 inches	2
24 inches	3
30 inches	4
36 inches	5
42 inches	6
48 inches	7
54 inches	8
60 inches	9
66 inches	10
72 inches	11

STEP 3: Next, solder the anode of the first LED to the wire coming from digital header position 0 (the 6-inch colored wire).

STEP 4: Solder the cathode of the first LED to the wire coming from the GND header.

STEP 5: Solder the remaining LED anodes to the wires from the digital header positions in order of their distance along the string: the second LED at 12 inches away attaches to position 1, the third at 18 inches away to position 2, and so on. When everything is wired up, you will have a 6-foot chain of LEDs attached to your Arduino controller, all ready for some programming.

PROGRAMMING THE LIGHTS

Since you have your Arduino board, you should also have the USB cable required to hook it up to your computer, and the free open-source software downloaded and installed on your computer. Time to fire up the software and plug your controller board in.

Following the instructions in the computer software, you can program your lights to do pretty much anything you want. I've included a bit of code below to start with (and if you don't want to transcribe this, I'll host the native Arduino code file on www.geek dadbook.com so you can download and install it instead).

```
void setup (){
for (int i=0; i < 12 ; i++){
pinMode (i, OUTPUT);
}
}
void loop (){
chaseLR();
chaseRL();
center_out();
odd_even ();
}
void chaseLR (){
/*This function does a Left to Right Chaser sequence
adjust the time_delay variable for longer or shorter delays*/
int time_delay = 100;
for (int i=0; i < 12; i++){
digitalWrite(i,HIGH);
delay(time_delay);
digitalWrite(i,LOW);
}
}
void chaseRL (){
/*This function does a Right to Left Chaser sequence
```

```
adjust the time_delay variable for longer or shorter delays*/
int time_delay = 100;
for (int i=11; i >= 0; i—){
digitalWrite(i,HIGH);
delay(time_delay);
digitalWrite(i,LOW);
}
}
void center_out (){
/*This function lights the lights starting at the center and going in
    both directions
adjust the time_delay variable for longer or shorter delays*/
int time_delay = 100;
for ( int x= 0; x < 7; x++){
int i = 6 + x;
int y = 6 -x;
digitalWrite(y,HIGH);
digitalWrite(i,HIGH);
delay(time_delay);
digitalWrite(y,LOW);
digitalWrite(i,LOW);
}
}
void odd_even (){
/*Alternates between the odd and even LEDs */
int time_delay = 500;
int count = 10;
for (int x = 0; x< 12; x++){
digitalWrite(x,LOW);
}
for (int i = 0; i < count; i++){
for (int x = 0; x< 12; x++){
digitalWrite(x,LOW);
}
for (int y = 0; y <12; y+=2){
digitalWrite(y,HIGH);
}
```

```
delay(time_delay);
for (int x = 0; x<12; x++){
digitalWrite(x,LOW);
}
for (int y = 1; y <12; y+=2){
digitalWrite(y,HIGH);
}
delay(time_delay);
}
}
```

With this code installed, you should be able to test your lights, using the power coming from the USB connection. This is your chance to troubleshoot any bad connections and fix them. When done, unmount it from your computer and hook it up with the AC power adapter and watch your lights twinkle!

MAKING THE ORIGAMI SHADES

All that's left is to make the shades to go over your bulbs. I've included the instructions (see diagram) for a simple origami box you can make in a variety of colors for a cheerful effect (thanks to www.wikihow.com for the basic info), but of course this part of the project is wide open to customization. Feel free to use anything from paper cups to Ping-Pong balls instead!

Once you have your shades made, slip them over the LEDs. You may have to use a little tape if the hole on top of your shade is too large.

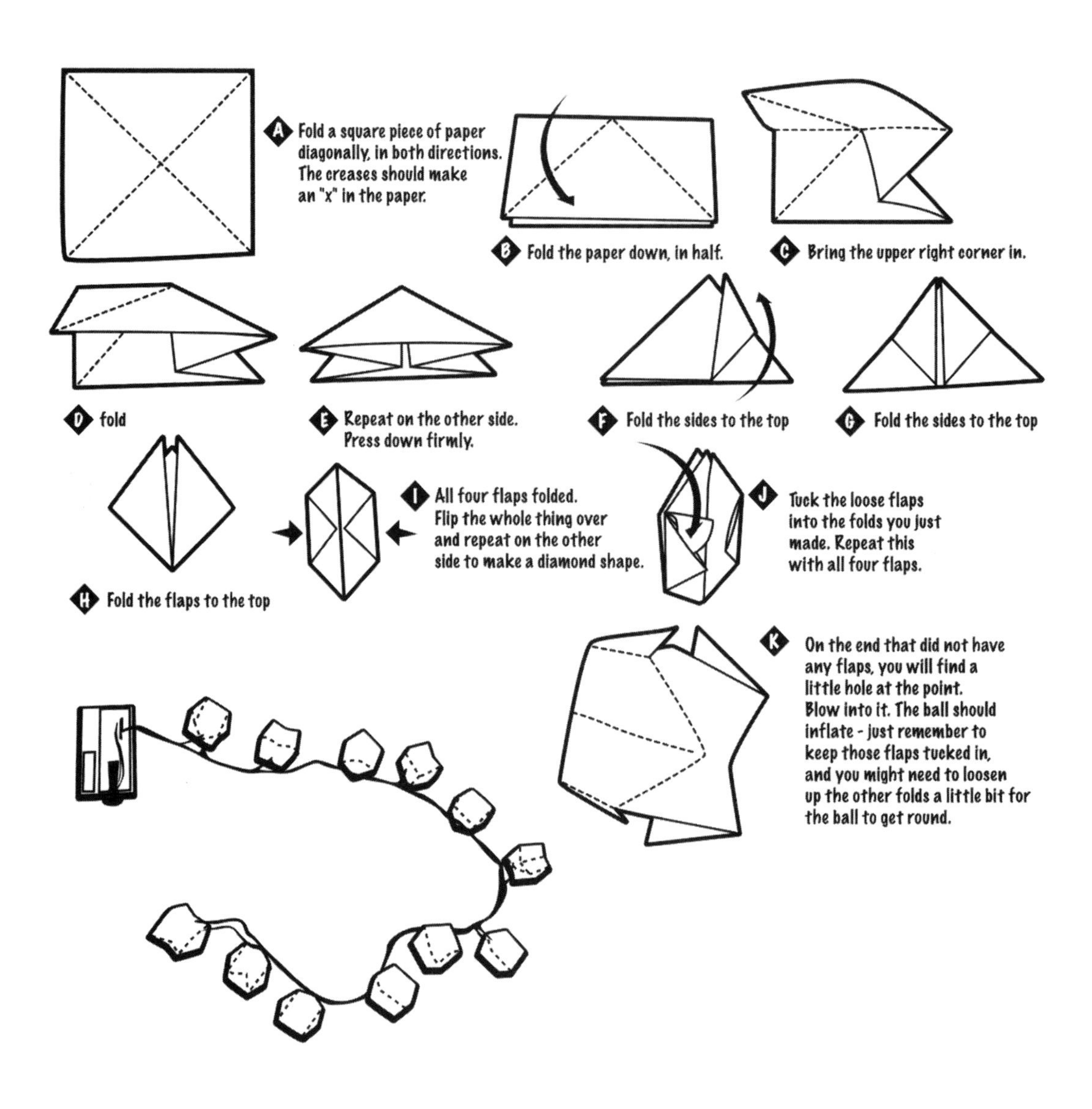
A Fold a square piece of paper diagonally, in both directions. The creases should make an "x" in the paper.
B Fold the paper down, in half.
C Bring the upper right corner in.
D fold
E Repeat on the other side. Press down firmly.
F Fold the sides to the top
G Fold the sides to the top
H Fold the flaps to the top
I All four flaps folded. Flip the whole thing over and repeat on the other side to make a diamond shape.
J Tuck the loose flaps into the folds you just made. Repeat this with all four flaps.
K On the end that did not have any flaps, you will find a little hole at the point. Blow into it. The ball should inflate - just remember to keep those flaps tucked in, and you might need to loosen up the other folds a little bit for the ball to get round.

FINISHING TOUCHES

And there you have it! You have a fully operational battle station . . . er, string of programmable lights. Of course, there are a lot of exposed wired and soldered connections, so you may well want to tidy things up for long-term presentation. Any combination of colored electrical tape, wire caps, and zip ties may bring things under control. And if you really want this to look like a hip project, mount your Arduino controller inside an Altoids tin (all the rage in the DIY electronics community), with a hole drilled into the side for the power adapter to run through.

And think, you only have to build about ten more of these to light up your Christmas tree!

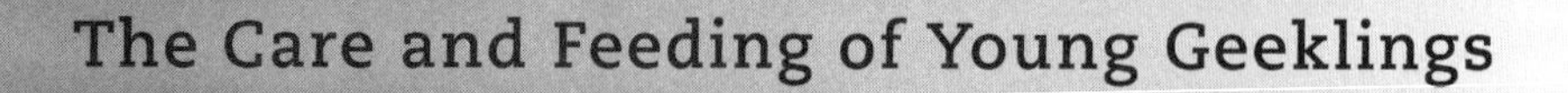

The Care and Feeding of Young Geeklings

by David Hewlett

David is a self-professed computer nerd, and a science fiction geek since childhood, which came in handy when he was cast as Dr. Rodney McKay on the TV show Stargate: SG-1, *a character that went on to be a main cast member on the spin-off series* Stargate: Atlantis. *You can find him online at www.dgeek.com, where his worshippers maintain an active community of friendly geeks, or follow his ongoing antics and his gushing commentary about his son, Baz, on Twitter, where he is dhewlett. And if you have even a tiny funny bone, check out his movie* A Dog's Breakfast, *a hilarious little black comedy he made with his wife, producer Jane Loughman, his sister Kate, and his dog Mars.*

My first response to the [first] Geek Dad book was a combination of joy and anticipation. My second, resentment and depression—as I realized my son was still too young to do any of the cool things suggested in it. Clearly, I was going to have to replace him with a more advanced model. But failing to get that strategy sanctioned by my wife, I resigned myself to searching for anything of geek interest that might be available to me at this tedious stage of development. At least until my newly natal nerd had downloaded and installed all the necessary upgrades for the good stuff.

Discussing children with geeks is like reciting the first 100 decimal places of pi to a hot young actress. Sure you sound cool doing it. But chances are, the only thing that they're going to understand is that your lips are moving. Not being a fabulously famous and successful star of film and television like myself, this is something

you'll probably have to take my word for (and if you are, can you introduce me to your agent?). Remember, "cute" may work on civilians, but nerds have an innate distrust of the stuff. To geeks, cute means Ewok. Enough said.

The term *baby* doesn't do anything for me either. There's nothing high-tech, edgy, or drool-worthy about a "baby." Besides, I already have my baby. It's a 17-inch quad core MacBook Pro. I'll use the term *life-form* instead.

Initially your new life-form (see how much better that sounds!) is going to look like a Gollum-eyed tree frog, and it won't want to do anything. By anything, I mean anything I want to do.

Everyone and anyone (me, for example) are going to have an opinion on how best to program and configure your recently shipped life-form. The most unsettling, when your offspring is of the male persuasion anyway, will be their unsolicited and highly detailed thoughts on circumcision. The key is to ignore them. If you have to take advice, take it from me, I've been on television. Think of parenting like directing a movie (yes, I've done that, too; aren't I amazing!). Be decisive. Make swift, sweeping, and often wildly inappropriate choices and make them loudly. Then change your mind (quietly) about them later.

Think of your new life-form as an unformatted hard drive. It is not going to be delivered wearing an I'M WITH GENIUS T-shirt complete with an arrow pointing up at its wide-eyed alien face. That arrow is pointing at you. Don't be disappointed. Be an EH (Early Humanoid) programmer! This is an opportunity to teach your very own Tabula Rasa that Tom Baker was the best Doctor Who and that the only Star Wars movies worth watching are the first three (made). Unhappy with "baby" bedtime books? Photo novels of *Close Encounters* and *Invasion of the Body Snatchers* (the one with Spock in it) and *The Thing* (Carpenter's remake, of course) are perfect nighttime reading. Your new life-form isn't going to know the difference. When I tell my son that I am his father, he launches into a very credible Luke Skywalker, holding one hand to his chest and screaming

"No" with real conviction. Making yourself happy is what being a great geek dad is all about.

There is a common belief that babies don't come with manuals. It's wrong! There are hundreds of thousands of baby manuals out there. I suggest that you buy each and every one of them. Sure, with minimal study, you can raise and nurture your child. But why stop there? A thorough understanding of the schematics and operating system that run your new life-form will allow you to put other parents to shame with their lack of trivial information about the process. Make it a point to read ahead. Familiarize yourself with the "system requirements" your child needs to perform various operations. You can then traumatize fellow parents by claiming that your life-form is capable of spectacular, wildly advanced (yet still believable) motor and comprehension skills. Like early adopting tech nerds, it drives parents crazy to think that their model may not be configured as well as yours. Just make sure you do your research. You don't want to look like an idiot by claiming your little genius is counting in binary when it's not yet able to support the weight of its own head.

Another added bonus to reading these manuals is the infinite number of new diseases and medical conditions you'll find to worry about. Baby manuals put the *hype* in *hypochondria*. Just make sure to couch your own fear of infection and injury in terms of concern for your life-form's well-being. This subtle distinction can transform you from whining wimp into a respected parental figure.

"Baby experts" will try to confuse you with terms like *soothing*. Do not be misled into thinking that there is anything "soothing" about getting your new life-form to stop screaming. The most effective technique we encountered to combat this torturous behavior seemed completely counterintuitive at first, but it turned out to be a Jedi-junior mind trick. The process consists of a slightly unsettling combination of tight swaddling, gentle shaking, and the most obtrusive-sounding shushing hiss directly into the new life-form's ear. The idea is to create a virtual prenatal environment. The charm of this technique is that it's basically telling you to do exactly what you feel like doing: bind, shake, and hiss at the screaming brat.

Of course it's very hard to soothe an agitated life-form if you don't know what it's complaining about. The key is that each distinct piercing screech needs to be directly associated with one of a handful of classic diagnostics: input (food), output (solids or liquids), reboot (sleep or lack thereof), and force quit (generally in the form of gas from one end or the other). I took this age-old hypothesis one step further and set out to write a brilliant program that would analyze and interpret each screech. I had begun a similar project with dog behavior before my dog went from pet to pest with the arrival of our child. By creating a comprehensive database of different cries and linking each to an array of documented solutions, I discovered that it was more accurate to flip a coin. This freed up more time to impress my wife with precise spreadsheets detailing each deposit made into our life-form's nappy. Nothing says Quality Dad like a poo pie chart.

A true geek father consumes whatever his new life-form just did. This is partially about making certain that whatever has been gobbled down is not poisonous. But the fact is there's no drama in sitting by helplessly wondering what this latest ingesting will bring. My son drank sunscreen. Concerned (and curious) about the effect those microscopic mirrors (which I'm pretty certain sunscreen is made up of) would have on his newly formed digestive tract, I promptly did the same. The result was a subtle but truly unpleasant feeling of nausea that lingers with me to this day. I can't even look at the sun without feeling queasy. That said, I am less susceptible to hot and spicy food—and am an heroic father figure.

This selfless practice can be traced back to my grandmother. As family legend would have it, at an early age (certainly pre-twenties, I should think), I shoved a marble deep up my nose. My family went to great lengths to dislodge that shiny, slimy obstruction, but to no avail. In desperation, my grandmother (obviously a geek pioneer) worked one of those little colored glass spheres into a similar position up her own nasal passage. It worked like a charm. We were both rushed to the emergency room, where we had the offending orbs removed by trained professionals.

You don't worry about the child's safety because you love the child. Children are a huge investment, one that a friend of mine once compared to giving birth to a ticking financial time bomb. Investments need to be protected.

Baby monitors are for losers. When else is your other half going to let you put a camera in a bedroom? This is a once-in-a-lifetime opportunity. Nerds like us, baby, we were born to run with it. Don't limit yourself to some lame consumer product. Our neighbor looked at her generic baby monitor screen one night to discover that her child was gone. Not only that, the room had been completely redecorated. If you haven't yet figured out that it was picking up another monitor, you should seriously reconsider breeding.

Why limit your surveillance to one bedroom when you can wire the entire house? In fact, push for a total property security system hardwired to a video monitoring station housed deep in a cement bunker or panic room. You never know. If your other half isn't buying it, go for the heartstrings. They grow up so fast, you don't want to miss a second of it, and twenty-four-hour surveillance means you never have to.

Now that the money is flowing, another purchase that you, as a geek, are going to agonize over is the car seat. Non-nerds see these as a simple safety requirement. But we will never forget the lesson learned from Jor-El as he jettisoned little Kal-El from his dying world. If you don't pimp your offspring's ride like a Kryptonian scientist father, how can you expect a Superman! (This is probably a good time to mention that exposing your pregnant partner or newborn to exotic forms of radiation or radioactive insects in the hopes of creating a super-powered masked avenger is a bad idea. The process is generally frowned upon, and radioactive experimentation may dramatically reduce your ability to have future spawn.) Your life-form's car seat should be thought of as a 007-like emergency escape pod. It could be deployed for numerous planet-saving reasons, many entirely unrelated to the minor fender benders brought on by radical hormonal changes in the female entity with whom you have chosen to procreate. The point being, why stop at ballistic nylon and multipoint air bags? What could be more secure than a fully armed and armored robotic attack vehicle? As I always say, the best defense is to be offensive.

The stroller is another geek favorite. Naturally, your instincts are going to point you toward a fully equipped cross-country assault vehicle. Ignore them. Sure it looks great online, but you don't want to be carrying a baby Batmobile around a crowded international terminal while your son is trying to board the plane via the luggage carousel. Go with something as light and portable as possible; you'll live longer.

As with computers and technology, the lesson with all important life-form-related purchases is simple: Buy the most expensive version you can get away with. The more money you spend, the better parent you are.

Since the closest thing to Dungeons and Dragons that these early humans are capable of is sucking the lead paint off your collectible knights and wizard figures, entertaining them can be tricky. Here's a couple of time killers that worked for me:

Curl your hand, palm up, fingers pressed together in a closed flowerlike shape. Make pleasant little chirping noises and begin to open and close the finger "petals" in a random pattern. Suddenly, open your hand fully and launch it at your life-form's face. Be careful to avoid damaging your offspring's ridiculously large eyes. Hold the child's face and make violent sucking motions. This is done by performing finger push-ups, pressing your palm against the struggling child's nose and mouth. All this should be accompanied by a terrifyingly high-pitched squeal. I call this maneuver the Face Hugger (for obvious reasons) and I knew I had a hit on my hands (quite literally) when I received that first toothless giggle. From the same Alien franchise comes another classic, best performed at mealtime. Begin by casually tucking an arm under your shirt. Pretend to choke and splutter on your food. This will draw the necessary attention for stage two. Launch into a fit of painful convulsions and mime full-body retching. Throw your head back and thrust your chest outward while punching out through the fabric of your shirt with the concealed hand. Remember to stretch that fabric to its limit for maximum effect (squeezing a handful of ripe raspberries through the cloth at the same time is a nice added touch that my son introduced me to). Once they get over the shock and emotional scarring of that initial attack, your life-form will be hooked.

I have merely scratched the surface of life with your nerdling. Let me leave you with one last suggestion. While the child is in full REM sleep, take a look at those little hands and feet. Take in that soft little toothless face. Consider the incredible potential contained in that tiny frame. There is nothing on earth as powerful and as beautiful as what you have created. Lean down and put an ear to that happy little kid's chest. Listen to the sound of that furiously beating heart, and know that there is no technology that can match it. Yet. Enjoy!

GAMING THE SYSTEM

Make Up Your Own Combat Card Game

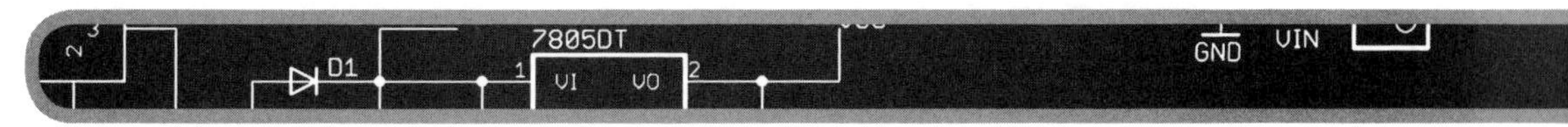

(Project Idea by "Z")

Combat card games, or CCGs, are hugely popular for their mix of imaginative RPG-like game play and the addictive collector aspect that goes along with them. Cards may represent characters, monsters, spells, or gadgets that interact with each other in interesting ways. Games often require an exciting mix of luck and strategy to win, and with many of the game systems, new cards are put out all the time, adding to the possible strategies (and to the things you need to buy to keep up). You also can't have just one deck of cards. There are booster decks and new specialty cards, and depending upon the style of game you'll be playing and your opponent's deck, you have to be able to put together just the right deck for yourself.

It's all a bit of carefully crafted commercialism, since the games are also always tied into animated television shows, game accessories, and other toys. But at their core, the games are fun, engaging play for geek parents and kids alike, demanding strategy, luck, memory, and math. Getting into these games with your kids, and supporting them playing with others, isn't the worst parenting decision you'd ever make.

But it could be one of the most expensive.

Packs of the cards don't come cheap, and there are new sets released all the time. On top of that, some cards are released in limited editions, which drive up prices. So, if you want to start playing, be ready to start paying. Or try the DIY route and make your own!

PROJECT	MAKE UP YOUR OWN COMBAT CARD GAME
CONCEPT	Customize a standard deck of playing cards into your own personal combat card game.
COST	$-$$
DIFFICULTY	⚙ — ⚙⚙⚙
DURATION	☼ — ☼☼
REUSABILITY	🌐🌐🌐🌐
TOOLS & MATERIALS	Standard deck of playing cards; stickers; markers; and creative supplies; suitable playing surface

This is a game for 2–4 players using one customized deck of playing cards. If you want to work with a larger group, add additional decks as needed. And if you have experience with CCG, feel free to expand or alter these rules to make the game more challenging and fun for yourselves.

STEP 1: Build the Deck

Take a standard deck of playing cards and remove the jokers (which may be used later for alternate rules variants). Now separate the face cards from the rest of the deck. Divvy up the face cards among your players and start customizing them.

By customizing, I mean draw on the cards, put stickers on them, do whatever you like to make them not just playing cards but COOL COMBAT CARDS. What's the idiom for your combat game? Is it robots? Dragons? Animals? Aliens? It's up to you what you put on the cards, just as long as you make sure the suit is still obvious and you don't alter the backs. These customized face cards will become the Character Cards, the warriors battling for supremacy on the Playing Surface.

STEP 2: **Set Up the Game**

Shuffle the Character Cards and deal two cards to each player, who should then place the cards faceup in a row in front of them. With each player's Character Cards in place, the remaining Character Cards and the Number Cards should be shuffled and placed in a common Draw Deck between the players. Each player should draw a single card from the Draw Deck, then the highest number card drawn will go first, with play progressing clockwise from that player. With the play order now decided, place the used cards facedown to one side of the Draw Deck to begin the Discard Pile.

STEP 3: **Play the Game**

Each round, a player selects a single card from the Draw Deck. The player then has four options:

Use an Attack Card: The player may use a red Number Card (heart or diamond) to attack an opponent's black Character Card (club, spade), or may use a black Number Card to attack a red Character Card, for damage points equal to the attacking Number Card's value. Each Character Card is understood to have 11 Hit Points, and any attached Attack Cards will be stacked faceup below the attacked Character Card to represent damage. After a Character Card has accrued 11 points of damage, its Hit Points have been exhausted, and the card is flipped over to denote that it is out of play (and any attached Attack Cards are placed facedown in the Discard Pile). If, at any time, a player has no Character Cards in play, s/he is out of the game.

Use a Buff Card: The player may use a red or black Number Card to buff (or heal) one of his own red or black Character Cards restoring Hit Points equal to the Number Card's value. Buffs may be applied only to damaged Character Cards, and recovered Hit Points—represented by the Buff Cards being stacked faceup above the Character Card—may never exceed the original 11. If a Character Card is taken out of play, these Buff Cards are also placed facedown in the Discard Pile.

Pass: The player may simply add the drawn card to his/her hand (with the face value concealed from other players), take no further action, and pass clockwise to the next player.

Call for Reinforcements: Should the player draw or have in hand a Character Card, s/he may place it faceup on the table adjacent to his/her other Character Cards, thus supplementing his/her defenses.

Play continues until only one player has in-play Character Cards remaining, and that player is declared the winner of the Playing Surface. Should game play exhaust the Draw Deck before a winner is determined, the cards from the Discard Pile may be shuffled and recycled as the new Draw Deck.

POINTS OF CLARIFICATION:

- The total Hit Points of a Character Card at any given time is equal to 11 minus the sum of the values of attached Attack Cards, plus the sum of the values of attached Buff Cards.
- Total Hit Points for a Character Card cannot exceed 11 in the standard game.
- Character Cards cannot be brought back into play in the standard game.

OPTIONAL RULES:

- Jokers may be included and used as Hyper Buffs capable of resurrecting out-of-play Character Cards. Once used, Jokers are treated just like Character Cards in that they are flipped facedown and not returned to the Discard Pile.
- For more advanced play, Character Cards may be healed only by matching suit Buff Cards rather than matching color.
- Alternately, matching suit Buff Cards may be allowed to heal a Character Card above the 11-point limit for a temporary time period (perhaps one or more rounds).

If you and your kids have some experience with how the traditional CCGs work, you can take this game even further, allowing for certain card types to do more damage than others. Or you could have special defense cards that each player can hide facedown under a Character Card, which will absorb damage from a specific Attack Card (only even-number hearts, odd-number clubs, prime numbers), or even trigger an out-of-turn attack on another player. The possibilities are as endless as your imaginations.

Fantasy Gaming Terrain

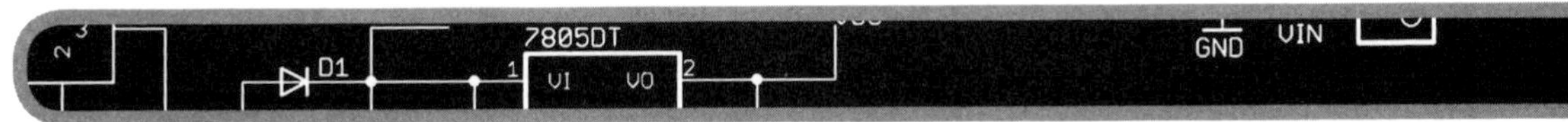

(Project Idea by Michael Harrison)

A lot of geeks are gamers, specifically fans of role-playing games like Dungeons and Dragons. And as we become parents, we take on an important new role: game master (aka dungeon master). If we can get our kids hooked on games like D&D, we help ensure their great geeky nature for life. And so we want to make it as fun for them as we can.

When I played, we usually used maps. Sometimes they were premade, but more often we used plastic-covered rollout mats that we could use dry-erase markers on to define spaces. We also hand-painted our lead figures. But today, the available materials have gotten much better. These days, you can get premade pieces of terrain like building blocks of castles, cutaways of dungeon corridors, or rocky hills and forests that can be laid on grids and used with scale figures. This helps you visualize the game even better and bring it to life in 3-D.

But, of course, these materials cost good money, and more often than not, you have to fit your ideas and the game you want to play to the available products, rather than the other way around. That doesn't have to be the case, though. It's actually rather easy, using materials similar to those we use elsewhere in this book for candy molds, to make your own terrain pieces to step your RPG gaming experience up a notch or two. With some modeling clay and silicone molding compound, you can make your own customized dungeon tiles for a lot less money, and have a great afternoon with your kids in the process.

PROJECT	FANTASY GAMING TERRAIN
CONCEPT	Make your own model terrain for fantasy gaming, using simple silicon molding compound.
COST	$-$$
DIFFICULTY	⚙⚙ – ⚙⚙⚙
DURATION	☼☼☼ – ☼☼☼☼
REUSABILITY	🌐🌐🌐🌐
TOOLS & MATERIALS	Polymer clay (Sculpey brand); silicone molding compound; plaster of Paris; clay shaping tools; latex house paint

This basic project teaches how to make basic tiles that look like the floors of a dungeon or castle, which you can use to assemble the floor plan of any such space you might need. Once you've mastered this technique, you'll be able to move on to other textures, shapes, and sizes that will enable you to create every part of a gaming scenario you might need.

STEP 1: Decide on a size and shape for your first tile mold. Because you'll be using these tiles in a modular fashion, think about creating a simple 2-inch-by-2-inch square piece. This will allow you to quickly build dungeons for your players on the fly.

STEP 2: Using a ruler, measure and shape your tile with the clay. Straighten the edges with the side of the ruler and continue to flatten the piece of clay until it is tile-shaped, a square 2 inches on each side, and about ¼ inch thick.

STEP 3: Most tactical role-playing games use 25mm scale miniatures and a 1-inch battle grid. With your clay shaping tools, slice into the top surface of the tile to create this grid. Add details to the tile to fit your theme; potholes, wood planks, or cracked cobblestones all add to the authenticity of your tile.

STEP 4: When you have a tile you like, cure the clay in your oven. For a ¼-inch tile, this usually takes about 15 minutes at 275 degrees.

STEP 5: After the tile has cooled completely, it is hard enough to take the mold. Combine enough silicon molding compound to surround the tile, then roll it into a ball.

STEP 6: Place the tile on a flat surface, finished side facing up. Gently press the compound into the center, spreading it around the entire tile until that side is covered. Tap the top to ensure that the compound fills all of the nooks and crannies of your masterpiece.

STEP 7: Let the compound cure. This will take around 20 to 30 minutes.

STEP 8: When the mold has set, slowly remove the clay tile from inside. Slow and steady, otherwise you'll tear the mold. Feel free to keep the clay tile on hand; you can paint it and use it just like the casts you'll make.

STEP 9: Now mix up some plaster. You'll want a batterlike consistency. Spoon it into your mold. Let it overflow slightly so that you know your mold is full.

STEP 10: Very carefully tap the sides of the mold to get out any bubbles.

STEP 11: Let set for about 30 minutes, then pop out your tile. Make enough for your vast and treacherous dungeon, and then just paint

them all with latex house paint to your desired colors (gray flagstones are always good).

You can create a mold for just about anything, so if you want to add walls to your dungeons, make some bricks and build them (regular Elmer's Glue works great to stick them together).

An Extra Geeky Idea

Don't stop with fantasy architecture: Try your hand at treasure chests, gelatinous cubes, or even futuristic spaceships. You can make molds out of real materials, like pebbles and stones, pieces of wood, or any other interesting texture you like, to add more dimension to your dungeons. In the end, it's all up to your imagination.

Now you can build your dungeon to match your ideas, rather than the other way around. Of course, if you need somewhere to start, you can check out some great premade molds at www.hirstarts.com.

Pun Wars

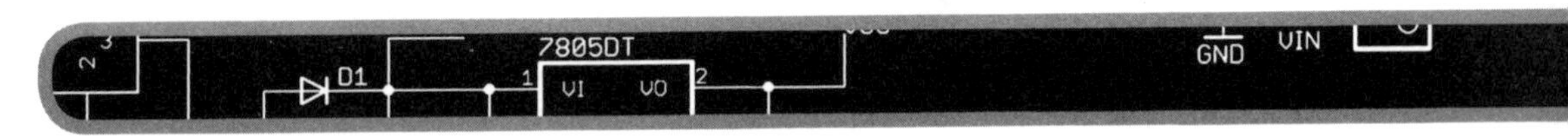

"The pun, or paronomasia, is a form of wordplay which exploits numerous meanings of a statement, allowing it to be understood in multiple ways for an intended humorous or rhetorical effect. These ambiguities can arise from the intentional use and abuse of homophonic, homographic, metonymic, or metaphorical language."

—WIKIPEDIA

The pun is also the most sublime—and often sublimely reviled—kind of humor. Sadly, I cannot explain why it's reviled. I love puns. Puns have a geeky cachet unlike any other kind of joke. When I was in high school, I would have pun wars (we called them our "Pun-ic Wars") with my friend Bill that could last an entire free period. We once made a girl cry laughing at us—er, WITH us.

Puns require a knowledge of subject matter both broad and deep, a quick mind, a savage wit, and an excellent vocabulary. Punning can be performed as a pair or a group and is great fun.

Which is why it's a great family activity as well!

PROJECT	PUN WARS
CONCEPT	Play a game of puns with your family.
COST	$
DIFFICULTY	1 – 4 (gear icons)
DURATION	1 – 4 (sun icons)
REUSABILITY	1 – 4 (globe icons)
TOOLS & MATERIALS	Absolutely none is needed. Pencil and paper can be handy if you want to play a scored game, and six-sided die for a challenge game.

Not long ago, our family was out with another family for dinner at our favorite Chinese restaurant, and one of the adults let slip the old standard pun about the knight of the round table who was so fond of steak: Sir Loin.

Things went south from there:

Knight who was always sure of himself	Sir Tain
Knight who was always angry with people	Sir Lee
Knight who did whatever he was told	Sir Vile
Knight who was fond of snakes	Sir Pant
Knight who was very lucky	Sir Indipity
Knight with a very round belly	Sir Cumference
Knight with a sharp sense of humor	Sir Donic
Knight from Eastern Europe	Sir Bian
Knight who painted with the Impressionists	Sir Zanne
Knight who also practiced medicine	Sir Geon

The best part was, it wasn't just the parents who were throwing out the puns above. The kids were into it, too, and doing their best to play. That's where the inspiration for this project came from.

But it also occurred to me that pun wars have a tremendous educational value. Beyond having the vocabulary and understanding of homonyms, the punner has to be able to come up with conceptual leaps in logic very quickly. It's almost like early training for debate, with a dash of absurdity and a heavy dose of silliness.

So how might one have family fun with puns, you ask? Well, you turn it into an organized word game. Pick a player to start, and pick a topic (you might come up with a number of topics beforehand, write them out on strips of paper, and then choose them from a hat or bowl). Player one says a pun ("I really think cooking shows on TV are very well done!"), and then the play passes to the left ("Really? I've always found them a bit half-baked"), and so on. Anyone who can't come up with a pun drops out, and the last person punning is the winner.

If you want to make it a more formal game, have each person get a point for each pun as the play goes around. Then a player can pass if s/he's lost inspiration for the moment but pick it back up later. Go for either a set number of rounds, or a set time, and whoever has the most points at the end wins.

An important point about how punning can play out: Whatever topic you start with can change over time. You may do a series of

puns on one topic and then turn a right corner based on part of the pun. For example, someone could tell the pun about the "Knight who was fond of snakes: Sir Pant," and the next player offers "Yes, and when he lost his job, he filed for Cobra coverage" (ooh, unemployment benefits humor!), and then, all of a sudden, you're off on Snake Names as a new topic. Which is how pun wars (for good or ill) can truly end up lasting for hours.

Obviously, the key with this game is that, when you're playing with kids, use topics that they'll be familiar with: superhero names, cartoon shows, sports teams, animals. Anything they will have enough connection to that they'll be able to riff on the names. After you play for a while, they'll become so facile with homonyms and word association, they will be the ones starting the pun battles. And then you will have won, because they'll be real geeks.

An Even Geekier Idea

You can play an ultra-challenging version with a six-sided die. On each turn, the player has to roll the die and state as many puns as the number they rolled. The player also earns a point for each pun, so the numbers can jump quickly. If someone is stumped, s/he can declare "punned out." If the play makes it back around to that player, s/he loses the number of points on a roll of the die. But if someone else also puns out, then a new topic is picked by the next player, and the play continues to a set time, or until all the topics are used. Most points wins.

Games to Play with Your Nano-Geek

Most of the projects in this book assume the kids you're dealing with have probably started preschool. But there's a very likely possibility that a decent percentage of you were given this book when your mutant offspring—er, your "kid(s)" are in the infant to toddler stage. You are a newly christened geeky parent in need, and I would never want you to be bereft of support.

Children are sponges. Especially young children. From infancy to preschool, all they're doing is eating, pooping, and learning. The scary thing is that they're learning from us.

When we talk to them, they're learning verbalization, the very idea of words and meaning, and then eventually communication and culture. And more so than any book-learning they'll do in school, what they're learning at this stage will form who they are and how they deal with the world.

If you want to raise your kids in your own geeky image, now's the time to start imprinting them with the language and culture of our tribe. To do so, you need to rethink how you speak to your pre-geek child. You need to wipe clean the cultural detritus of "normal" culture and replace it with the geek culture equivalents. All the little baby talk, silly poems, and simplistic games we play with babies and toddlers are artifacts of the culture that used to spurn us, so why would we perpetrate it upon the Next Generation (sorry, can't help capitalizing that)? No, we need to be like the Romans: Absorb the old, primitive culture but recast it in line with our advanced ways.

To help you achieve this, I realized I need to provide you with some ideas. And where do the best ideas come from these days? Why, from Twitter, of course!

I sent out the following message on my Twitter account:

@fitzwillie: Hey, everyone! I need ideas for geeking-up the silly games we play w/infants-to-preschoolers.

In response, I received quite a few wonderful ideas, which I will now share with you.

"THIS LITTLE PIGGY"

The classic toe-counting/foot-tickling game has been a favorite for giggling babies since the eighteenth century. The commonly accepted version goes like this:

This little piggy went to market.
This little piggy stayed home.
This little piggy had roast beef,
This little piggy had none.
And this little piggy cried "Wee! Wee! Wee!" all the way home.

The first geeky variant for this came from Xia Harris/@inbarati:

Taught my godson to play "this little broken robot" instead of "this little piggy"

Sounds like fun! Though I'd suggest just "robot" to match the meter of the original. As such, here's one possible Geek Dad variant:

This little robot went to Mars.
This little robot stayed home.
This little robot had a recharge,
This little piggy had none.
And this little robot cried "Bleep! Bleep! Bleep!" all the way home.

One other person had a great alternate for "Piggy." My friend Bill Moore/@thebillmo thought a *Lord of the Rings* version was a good idea:

This little hobbit went to Mordor.
This little wizard did roam.
This little dwarf had roast mutton.
This little elf had none.
And this little ranger cried, "Oi! Oi! Oi!" all the way back to the throne.

"PEEKABOO"

Of course, there's nothing more fun to do than spend hours with a baby, playing peekaboo and getting the same wonderful giggle reaction and look of surprise EVERY SINGLE TIME.

Peekaboo is a game thought by child psychologists to help children develop a sense of object permanence. So it's an awesome opportunity to combine critical early learning with your child's geek education. Here are some of the ideas I received in reaction to my call for help:

> "Peekaboo" becomes "Cloaked" and "Uncloaked"—Brian McLaughlin (on Facebook)

This is a reference of course to the science-fictional technology used in the various *Star Trek* shows that allowed spaceships to become invisible. Of course, at some point, your child will grow up enough to respond by scanning for residual tachyon emissions and know that you're there even when your face is covered.

OTHER IDEAS:

> I encourage my 1 yr old to open the docking bay for the shuttles during meals.—Darin Ramsey/@therealdarin

Ahh, feeding time at the high chair. What more blissful and pleasant time of day is there? Obviously using "the train's coming into the station" as the means to entice taking a spoonful of food must be updated!

Instead of "Simon Says," how about "Make it so, Number one!"—
Dan Lewis/@DanDotLewis

Again, another great *Star Trek* idea. In this case, in *The Next Generation* television series, one of Captain Picard's catchphrases is to confirm a command by saying "Make it so." Number One was his executive officer, Commander William Riker, who often was on the commandee end of an order. So, instead of the mythical Simon telling folks what to do, we replace him with the best captain of the starship there ever was (unless you think that was Kirk. Or Pike.)

Turn "red light, green light" into "Don't wake the Dragon."—
Kevin Wilson/@DocDraconis

This one has the advantage of wonderful simplicity. Instead of "green light," the it-person (really, the dragon) says, "I'm asleep." And instead of "red light," he just turns and gives a great roar, signaling the dragon is awake. Anyone who doesn't freeze has been eaten by the dragon. So much more fun!

Geekmom.com editor Natania Barron passed along the idea to change the classic circle-tag game "Duck, Duck, Goose" into "Droid, Droid, Wookiee." When the "Wookie" is picked, the children have to run around making a Wookie cry while trying to catch the picker before he gets back to the open seat. My friend Paulette Moore suggested a *Battlestar Galactica* version: "Human, Human, Toaster!"

But of course these are only a few ideas to get you started. Your geeky passions may differ slightly, and so the variations you teach your geeklings should as well. If you come up with any more, please post them on www.geekdadbook.com!

High-Tech Treasure Hunt

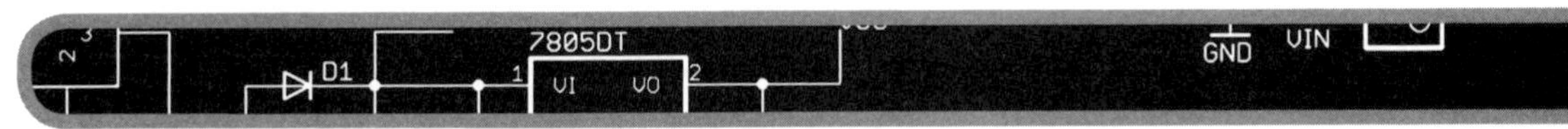

(Project Idea by Matt Blum)

Kids love adventure, and they love solving puzzles. They also love imagining themselves as the Intrepid Archeologist or Super Spy or World's Greatest Detective. Which is why treasure hunts are so much fun.

When I was growing up, my parents would often design treasure hunts for me on Easter mornings, leading me on a chase from clue to clue to find my candy-filled Easter basket (often hidden in the dishwasher—somewhere I'd never go looking). Every year since my kids have been old enough, my wife and I have done the same thing for them. It's become both a tradition and something the boys love and look forward to.

And over time, we've tried to become more sophisticated with the clues themselves, and the method of delivery. We've done video clues and left clues on computers and such. But it seemed like there had to be a way to step it up a level, considering how much technology there is in our home.

SIDE NOTE: when I was just out of college, my buddy Randall proposed a quantifiable scale for how technologically sophisticated one is. Just count the number of lasers in your home. At the time, CDs had finally supplanted tapes as the primary medium for music, and CD-ROMs were how you accessed rich media on your computer, so you could pretty much count lasers by how many of these drives you had. But it seems we've moved past that scale, since anyone can buy a pocket key chain laser these days (can you imagine how that would sound to someone in the fifties?). My thought now is that we can measure the technological sophistication of a household by the number of Wi-Fi-enabled devices there are. Smartphones, computers of all shapes (PCs, laptops, netbooks, iPads), TVs and BLU-ray players, set top boxes, game consoles, and more take advantage of home networking and access to the Internet. Count up these devices in your home and see what your score is. Mine (for today at least) is 15.

It occurred to me that we, like a lot of geek families, I'm sure, have a bunch of old smartphones and an iPod Touch sitting around not doing much. It also occurred to me that virtually every geek family has a wireless network in their house, to which those smartphones and similar devices can easily attach.

And thus the idea for a new kind of treasure hunt was born.

PROJECT	HIGH-TECH TREASURE HUNT
CONCEPT	Set up a treasure hunt in you house using Web-enabled devices on a local area network.
COST	$ (assuming you use tech you already own)
DIFFICULTY	⚙⚙⚙ — ⚙⚙⚙⚙
DURATION	⏱⏱⏱ — ⏱⏱⏱⏱
REUSABILITY	🌐🌐🌐 — 🌐🌐🌐🌐
TOOLS & MATERIALS	Home computer; home network; Web-enabled portable devices

Opening caveat: Out of all the projects in this book, this is the one you actually do yourself rather than doing it with your kids (at least the setup). That being said, it's going to be so much fun after it's done, I don't think they're going to mind.

In this treasure hunt, you'll mix Web-enabled tools with good old-fashioned sleuthing. Using one or more computers or smart phones, your kids will solve puzzles or figure out clues to take them to places where you've hidden secret words. Each secret word will unlock a new puzzle or clue and keep them moving until they finish the search.

STEP 1: Create a series of clues, each one leading to the next, as if you were doing a "normal" treasure hunt. They can be written riddle-type clues, maybe math or history related (just to put a little school learning into the fun), video clues of any sort, audio clues, and of course several physical clues somewhere in or around your house. Make a list of these, being sure the order is clear and that you know exactly what answer you're looking for and how flexible you want to be with that answer (with older kids, for example, you might want to make capitalization matter; with younger kids, you might list some almost-right spellings that you'd accept).

And figure out special locations or items around your house that will be good hiding places for secret words. The basic framework here is to have your kids start on a computer or phone and access a starting Web page for the hunt that lives on your local network. By answering a question correctly, they will be taken to another page. Perhaps that one gives them a clue to a place where they will find a secret word. They have to figure out the clue and go find the secret word. Enter that into the page, and they'll be taken to the next question or clue, and so on. Map it all out on paper first.

STEP 2: Download and install Web server software on your primary home computer—preferably a desktop. The easiest one to use is Apache, since the install is dead simple and it's free:

http://httpd.apache.org/download.cgi

I recommend setting it up as manual-starting, and on port 8080. Yes, these are geeky technical terms. If you are a reasonably computer-savvy geek, you should be able to pick up the next few steps pretty easily. If you're not, you may want to enlist the help of someone who is. These steps aren't all that difficult, but the terminology is pretty specific and could be confusing for the uninitiated.

STEP 3: Go to the htdocs folder of your new Apache install and create the files you will need—basically, a Web page for each correct answer, which will contain the next question on it as well. Start by modifying the index.html with the opening of your treasure hunt and a starting question, and then create each subsequent html file with the filename format <previous clue answer>.html. This makes it so that when the correct answer is entered in the previous page, the next page will be opened. It also makes it impossible to guess the next filename without figuring out the clues—obviously, if you named them "index2.html" and "index3.html" and such, anyone would be able to figure out how to skip steps.

STEP 4: Make any videos you will need, in QuickTime format—preferably MP4. Do NOT use HD video unless the only devices that are going to be used are iPhone 4s or iPads or something else that you know for sure can display it. I recorded a sample with my webcam, which worked perfectly. You can also create graphics—maybe a map or something similar. Drop every media file you need in the htdocs folder along with the html files, and embed them in the html. In one of the example scripts included below, a QuickTime video is embedded, which works great on Apple devices—it'll show a gray video box on the page, which you can tap on—this will pop up the video and, once it's played, return you to the page. Embedding images is pretty easy, of course, and there are tutorials galore online to help you if you don't know how.

STEP 5: The very basic JavaScript in the sample code below will take you to the next page if you enter the correct answer, or pop up a message if you enter an incorrect one. The programming is extremely straightforward and should be tweakable by nearly any geek or a friend. There are also, of course, tutorials and reference guides to Javascript online. If you're going to use a smart phone or similar device that has the ability to display source code, you may want to monitor the hunt, since the answer is right there in the code, and smart geeky kids might figure that out. Fortunately, Apple devices do not have this capability.

EXAMPLE CODING FOR THE PAGES

Below is the starting page. You can set whatever device is going to be the main tool for the hunt to default to this page when the browser is opened. If the treasure hunters are going to use a device other than the one from where the game is served (that is, has all the files on its hard drive), you'll have to set said device to find the original on the network. For example, since my desktop machine is named kermit, the net address of the starter file would be http://kermit:8080/.

In this example, the clue is "You'll have to be sharp to come up with the name of this sea creature!" The correct answer is "swordfish." If the word is entered in the black field provided, it will open the next web page.

Index.html

```
<html>
<head>
<title>Super-Secret Spy Web</title>
<script type="text/javascript" language="Javascript">
function solve(f) {
if (f.answer.value.toLowerCase() == 'swordfish')
```

```
window.location = 'swordfish.html';
else
document.getElementById('error').innerHTML = 'Sorry! Try again.';
}
</script>
</head>
<body>
```

You'll have to be sharp to come up with the name of this sea creature!

```
<br>
<span id="error" style="color:red"></span><br>
<form name="answerForm" onSubmit="solve(this); return false;">
```

Enter answer: <input name="answer" id="answer" type="text" size="10" maxlength="20"/>

```
</form>
</body>
</html>
```

This next page is named for the correct answer in the clue above, and asks the next clue. If the correct answer is given, it moves the treasure hunters on to the next clue.

Swordfish.html

```
<html>
<head>
<title>Super-Secret Spy Web—Part 2</title>
<script type="text/javascript" language="Javascript">
function solve(f) {
if (f.answer.value.toLowerCase() == 'opensesame' || f.answer.value
    .toLowerCase() == 'open sesame')
window.location = 'opensesame.html';
else
```

```
document.getElementById('error').innerHTML = 'Sorry! Try again.';
}
</script>
</head>
<body>
What Ali Baba had to say to get into the thieves' cave.<br>
<span id="error" style="color:red"></span><br>
<form name="answerForm" onSubmit="solve(this); return false;">
Enter answer: <input name="answer" type="text" size="10"
    maxlength="20"/>
</form>
</body>
</html>
```

This script shows how to embed a video file in a clue page. Remember, all the filenames and titles are up to you. Mine are just suggestions.

Opensesame.html

```
<html>
<head>
<title>Super-Secret Spy Web—Part 3</title>
<script type="text/javascript" language="Javascript">
function solve(f) {
if (f.answer.value.toLowerCase() == 'swordfish')
window.location = 'swordfish.html';
else
document.getElementById('error').innerHTML = 'Sorry! Try again.';
}
</script>
</head>
<body>
<object width="320" height="256" classid="clsid:02BF25D5-8C17-
    4B23-BC80-D3488ABDDC6B" codebase="http://www.apple
    .com/qtactivex/qtplugin.cab">
<param name="href" value="opensesame.mp4" />
<param name="target" value="myself" />
```

```
<param name="controller" value="false" />
<param name="autoplay" value="false" />
<param name="scale" value="aspect" />
<embed width="320" height="256" type="video/quicktime"
     pluginspage="http://www.apple.com/quicktime/download/"
href="opensesame.mp4"
target="myself"
controller="false"
autoplay="false"
scale="aspect"> </embed>
</object>
<br>
<span id="error" style="color:red"></span><br>
<form name="answerForm" onSubmit="solve(this); return false;">
Enter answer: <input name="answer" id="answer" type="text"
     size="10" maxlength="20"/>
</form>
</body>
</html>
```

Be sure to test the whole thing ahead of time, and BE SURE you clear the history and cache of each device you use for testing, or a reasonably smart kid will be able to find the URLs of the later bits without solving the clues—on Apple devices, Safari's cache is cleared via Settings -> Safari.

If you want to, and know what you're doing, you can even go to your router's control panel and restrict the devices being used so the hunters can't access anything outside your local network—thus preventing kids from using Google or Wikipedia to help find answers. Or you might want to prevent them from accessing anything but Wikipedia, or something like that. Every router should have the ability to regulate this. If not, and you really want to restrict them from accessing the Internet, you can always disconnect the router from the Internet—of course, that locks you out, too, but them's the breaks.

Pokémon Bingo

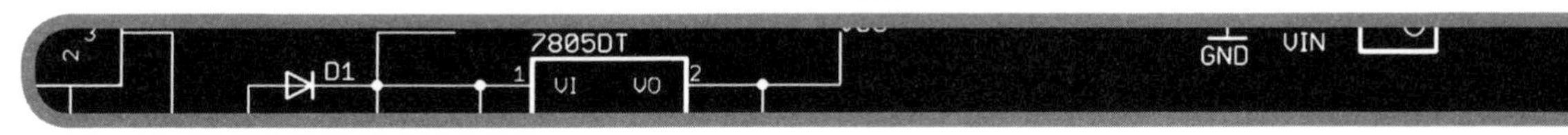

(Project Idea by Corrina Lawson)

We geeks worship at the altar of our entertainment media. We absorb, retain, and regurgitate obscure details from the shows and movies we watch—not unlike medical students are expected to do with anatomy and diseases. Except we enjoy it. And don't get paid as well for doing it.

Beyond picking up useless trivia, though, we learn. How do I know that iron and copper have similar properties? *Star Trek*! Or the fact that humans are mostly made up of water? I learned that from the old *Batman* TV movie where they desiccate that UN ambassador. And I can't even say how many science-y facts I picked up from MacGyver!

Because of our geeky tendencies, perhaps we're less wary of letting our kids watch too much television than the average parent. But there's a way to use our kids' favorite cartoon characters and shows to help them learn away from the TV set, too. We can turn them into the stars of games that require a higher level of attention and interaction. This idea to make learning more fun is the inspiration for Pokémon Bingo!

PROJECT	POKÉMON BINGO
CONCEPT	Make bingo cards based on the characters and creatures in the cartoons your kids watch.
COST	$
DIFFICULTY	⚙ — ⚙⚙
DURATION	☼ — ☼☼
REUSABILITY	🌐🌐🌐🌐
TOOLS & MATERIALS	Construction/printer paper; cardboard backing; white glue or a glue stick; drawings or printed images of animals; poker chips or pennies; willingness to watch what might seem an incomprehensible show

On the surface, Pokémon and similar anime series may seem to parents to be nothing but loosely scripted battles with strange incomprehensible creatures, all meant to make your kids demand the latest merchandise the next time you walk into the toy section at Target. At some point, you may want to throw up your hands and say, "I don't speak Pokémon."

But you do.

Pokémon and many other anime shows are invariably based on something that's very familiar. The Egyptian gods were the inspiration for the characters in various incarnations of the long-running show *Yu-Gi-Oh!*. The creatures used in battles on that show also were based on horses, tigers, and elephants. Another show, *Bakugan*, features dueling monsters, but many of those monsters are based on animals, insects, or mythological beasts. In *Pokémon*'s case, the characters are real-life animals that are twisted just a little (electric chipmunks) or animals that have been combined (like tiger-bats).

This means that if you are an enterprising parent, you can change the experience of watching these shows from a passive one into an interactive, geeky activity requiring a mix of active participation and deep learning of the associated subject matter. But it's going to take a little bit of dedication on your part to make this work: First of all, you're going to have to watch the shows!

STEP 1: Research

First, you must watch several episodes with your child (if you're not already doing this, of course). As the creatures appear, write down the animal that inspired it. For example, the most popular Pokémon, cute little Pikachu, is based on a mouse and has electric powers. So it's essentially an electric eel crossed with a rodent. Meowth is, of course, a cat. Charmander is a fire lizard, Tauros is a bull, Treecko is a gekko, Taillow is a colorful bird, Zubat is a bat, and Piplup is a penguin. You get the idea.

Because there are nearly five hundred named Pokémon, you may want to restrict your choices a bit and try to use the kinds of animals or creatures common to the more familiar characters on multiple bingo cards (in just the way the same numbers are used on real multiple bingo cards). Wikipedia has exhaustive lists of them, as do the plethora of Pokémon-related fan sites on the Internet. Bird- or cat-related creatures are plentiful in the Pokémon bestiary, intelligent plasma balls less so, making the selection of hybrid animals for your cards a relatively easy task. It might be handy to start by making your own matrix list with animal types on one axis (cats, dogs, lizards, rats, birds, snakes, and so on) and a box to check off each time you see that animal in a Pokémon creature. And yeah, I could have included all that research here in the project, but then you wouldn't get to learn it TOGETHER with your kids!

STEP 2: Make the Bingo Cards

Now that you have a list of animals, you can make up the cards for a bingo game. The kids should get involved with this as well, so everyone has ownership of the game.

A bingo card can be made simply with construction paper, using

rulers and pencils to create the squares. Or colored pencils or crayons can be used to draw the real-life animals that inspire the fantastical creatures.

For the cards, instead of the traditional bingo card on which there are letter/number combinations in the boxes, put the type of animal or creature in the box (either by name, a drawn/printed picture, or both). You should probably limit the cards to a 4x4 grid of animals so it won't be too hard to score a bingo in the period of one or two episodes.

Once you have the grid and drawings done, you can glue the paper to the cardboard backing to make it more durable. If you have access to a laminator, this could make the game pieces even that much more reusable.

If you or your kids are more computer-inclined than artistically inclined, look up the animals and print out images for the cards instead. There are also likely computer programs with which you could create the whole card online, but since the idea is to get kids away from staring at a computer or a television, the tactile method is recommended.

Just being involved in the process of making up the cards will get your kids thinking more about the animals that inspired their favorite characters. It will spur great conversation, and maybe they'll go off on tangents, wondering if there were any real-life lizards that actually spit fire (sadly not, though spitting poison is a decent alternative).

STEP 3: Play the Game

You will want to have at least one bingo card per kid, more if you have enough animals/creatures from your research. Because there may be only a few Pokémon in any given episode, try not to make getting four in a row too hard; use the real-life counterparts for a popular Pokémon like Pikachu, or an animal that is used as a basis for more than one creature, like a cat or a bird. That way, the kids will be able to cover as many squares as possible per episode and get bingos more often (which is more fun!).

Part of the fun and education will be in watching the show with your kids and playing the game with them. Encourage them to call out the animals as they see them, and if there are any disputes as to what the animal part of a given Pokémon is, you be the judge (since you did the research).

The poker chips or pennies can be used as markers on the bingo cards. Or a slightly more mercenary variation could be played with many pennies. Each player starts with a number of pennies equal to the total number of squares on all the cards being played. Each time one player scores a bingo, s/he wins all the pennies on the cards of the other players.

Once the cards are completed, keep them and the chips/penny jar next to the television. Whenever the kids are watching their favorite show, get out the game and urge them to play. If you provide incentives for winning bingo, such as a new Pokémon card or other treat, they'll remember better.

If the kids really like this activity, you can keep note cards or a small notebook next to the television so the child can write down more animals to create new bingo cards. And remember that this game doesn't work just with *Pokémon*—there are many other shows to which this model can be applied.

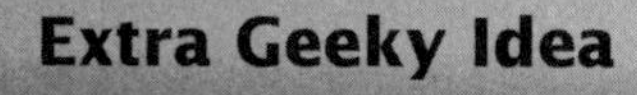

Extra Geeky Idea

The next time you take your kids to your local zoo, take the cards with you and play the game (maybe with nonpermanent markers instead of the chips/pennies). It'll be an awesome way for your kids to make the connection between the creatures in the shows they watch and the real animals if they can walk through the zoo, marking down on their cards "Ooh, there's a bat just like Zubat! There's a frog just like Bulbasaur! BINGO!"

This project will help your child see beyond the show as entertainment and translate it into a way to learn more about the real world around them.

In the process, you may also learn to speak Pokémon! Maybe.

Backyard Zip Line

Aka "The Zip Line of One Million YouTube Views"

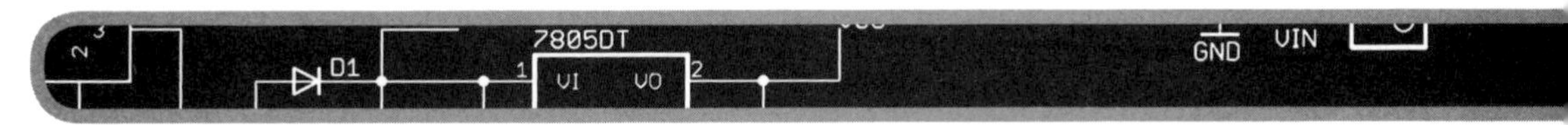

Celebrity Geek Dad Project by Jamie Grove

Jamie Grove holds the title "Director of Evil Schemes and Nefarious Plans" for the online merchant www.thinkgeek.com, perhaps the best source for geeky paraphernalia for discerning geeks everywhere. But Jamie doesn't just work there: He's a customer. Meaning he's a big helping of geek (and Dad) himself.

One day while dodging "red-hot plasma charges" (e.g., acorns thrown by my two rambunctious boys), I got to thinking about escaping on the greatest zip line in all of science fiction: the monoline from Jack Vance's 1957 classic *Big Planet*. Here's how it is described in the book:

"Wind blew in sails, and trolley wheels whispered down the monoline—a half inch strand of white Swamp Island cable. From the dome at Swamp City, the line led from spine to spine across three miles of swamp to a rocky headland; it crossed over the rotten basalt with only six inches to spare, swung in a wide curve to the southeast. At fifty foot intervals, L-brackets mounted to poles supported the line, so designed that the trolleys slid across with only a tremor and slight thud of contact."

As I daydreamed about Vance's amazing creation, I had a vision of setting up a sail-driven zip line. That would be so cool! However, after getting conked on the dome by three or four acorns, I came to my senses and figured it would be better to start with something a little less powerful but still geeky. And so I came up with a great new idea that was still pretty cool: a video-camera-enabled zip line.

PROJECT	BACKYARD ZIP LINE
CONCEPT	Build a zip line with a camera mount so that your kids can start earning money for college via YouTube.
COST	$$$$ (about $300)
DIFFICULTY	⚙⚙⚙⚙
DURATION	☼☼☼
REUSABILITY	🌐🌐🌐🌐🌐
TOOLS & MATERIALS	¼-inch aircraft-grade galvanized and uncoated cable; ¼-inch duplex cable sleeves; heavy chain; 12-inch jaw end galvanized turnbuckle; 8-foot 1x6 and scrap (18 inches); 2x2 pressure-treated lumber; 20 3-inch galvanized deck screws; double pulley trolley; large carabiner; ¼-inch eyebolt with nut and washer; ¼"-20 x 4" Screw; 2 ¼-inch-by-1¼-inch flat washers; 3 ¼"-20 wing nuts; square-style U-bolt; saw; a drill

The zip line is a simple concept. Begin with a cable stretched between two trees at high tension, add a pulley, and attach a means of holding on to the pulley. Then hope that you've given the thing a decent enough slope to move the mass of a child and or parent or womp rat along to the destination, but not so much slope as to cause said child or parent or womp rat to become part of the tree at the end of the line.

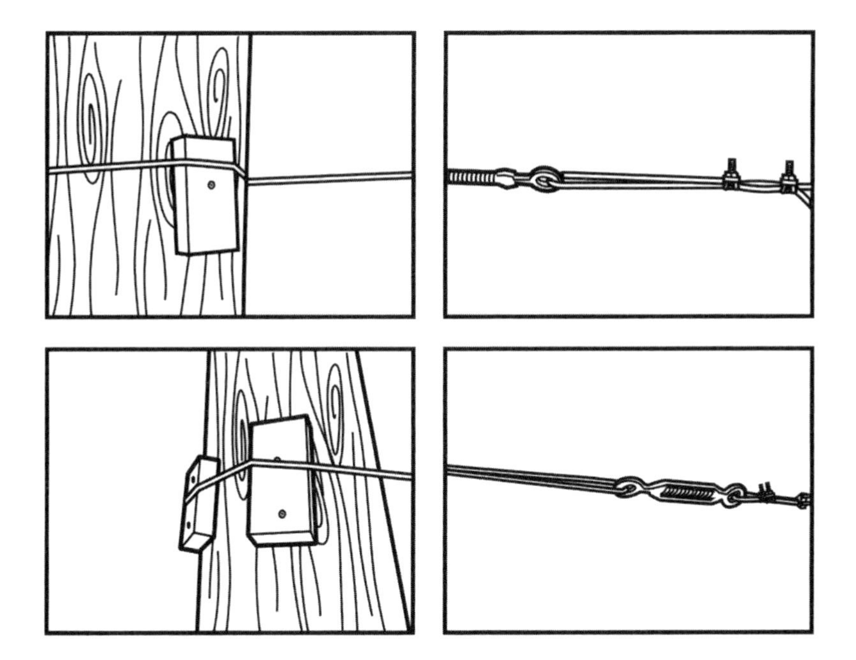

STEP 1: Select the Right Cable

Before I went to the store, I'd read plenty of articles talking about using $\frac{1}{4}$-inch aircraft-grade galvanized and uncoated cable. I scoffed at this, mostly because it seemed a bit excessive. After all, $\frac{1}{4}$-inch aircraft-grade galvanized and uncoated cable is rated to carry about 7,000 pounds. I've eaten a lot of pizza lately, but things haven't gotten quite that bad. Don't give into fiscal guilt at the cost of the stronger cable. Get the big stuff. Otherwise, only your smallest child will be able to use it, and you'll always be wondering if it's about to snap.

You will also need to get a pair of crimpers and duplex sleeves to close off the ends of the cable. Make sure the sleeves are rated for the $\frac{1}{4}$-inch aircraft-grade galvanized and uncoated cable you are going to buy.

STEP 2: Buy a Trolley

You might be tempted to try out your project with a simple household pulley rated for a weight well above that of your child. Don't. The worst part about using a simple pulley is that they are about a thousand times slower than a proper trolley (an exaggeration perhaps, but bottom line is they are less fun). The second worst part is that they tend to break and cause you to take your kids to the emergency room. So, without a doubt, you've got to go with a trolley. It will be the most expensive piece in your zip line, but it is worth it.

For my trolley, I selected a double-tandem pulley, which provides a better distribution of weight on the cable (and thus makes you go much faster by reducing friction). To be specific, I chose a Tandem Speed double pulley from PETZL because you can use it in two different orientations and it provides extra connections for carabiners. In addition, this trolley is rated for Tyrolean-size cables, which means I get to say "Tyrolean" every time someone asks me about the project.

For the handle, I used a piece of 2x2 deck railing. I drilled a ¼-inch hole in the center and attached the eyebolt with a nut and a washer. To attach the two, I used a large carabiner.

STEP 3: Figure Out What You Need Before Going to the Store

Of course you're excited; who wouldn't be? Well, before you run out to the store, do yourself a favor and make sure you know what you need to get.

Measure the length of the run

In my case, I have a lot of trees in my backyard, and my boys want several lines. So I built a test zip line of 30 feet.

Measure the circumference of the anchor tree.

The anchor tree is the tree at the BOTTOM of the run. This is the tree you'll be winding some chain around. I'll explain later why it should be the tree at the bottom of the run.

Determine the end height based on the length of the line.

For a typical (and safe) zip line, you should plan for a 6 percent rise over each 100 feet of distance. Also, depending on the weight and length, you should plan for a sag of about 2 percent over the same distance. Since our zip line is going to be 30 feet long, the end should be about 2.4 feet higher than the anchor (i.e., 30*.08).

Determine how much cable you really need.

Since we're going to build more than one zip line, I'm going to buy a big roll of cable. However, unless you are planning to set up a high-ropes course in your backyard, you probably don't need that much. So you get to teach your kids a little math here using our old friend the Pythagorean theorem ($a^2 + b^2 = c^2$).

A = 2.4 feet

B = 30 feet

C = square root of 2.4^2+30^2 or um . . . 30.09584 feet

All right, we could have just said 30 feet plus a bit more to go around the anchor tree, but we're trying to teach a little math to the kids, right? If they complain this is too simple, feel free to move from Euclidean geometry to the finer points of trigonometric functions.

In my case, about 50 feet of cable should be enough.

The anchor chain

You could get away with buying some extra cable and wrapping that around the tree to serve as the anchor, but I recommend using a chain. Chain is easier to link up to the turnbuckle and it is much easier to use the come-along with a chain (as my knuckles found out on my test line).

The turnbuckle

A turnbuckle is a nifty device that you'll see anywhere there are cables under tension. In essence, it's two long screws connected with a threaded buckle. The buckle is designed in such a way that turning it one way tightens (or loosens) both screws in unison.

Get one that has two jaw-end fittings and not one with eyebolts.

I would also recommend getting one that is at least 8 inches long. A 12-inch-long turnbuckle is probably better, because the longer the turnbuckle, the greater your margin for error when you cut the cable and hook things up.

The come-along

The come-along, or cable puller, is basically a hand-operated winch. You attach one end to a stationary object and unwind the cable so you can attach the other end to something you want to "come along."

In case you think you can do without this piece of equipment, I urge you to reconsider. I thought the same thing on my first trip to the store, and after the second trip, I wished I hadn't listened to myself.

STEP 4: Protecting the Trees

Unless you're prepared to drill a hole all the way through your tree, you'll be running cables and chains around the trunk and then putting them under tension. This is going to hurt the trees, and so will drilling a hole all the way through the tree. Let's not hurt the tree, shall we?

To protect the tree, buy one pressure-treated 8-inch 1x6 and cut it into blocks about 6 to 8 inches long. When you secure the blocks on the tree, use a pair of galvanized decking screws, 4 to 6 per tree. Yes, the screws will hurt the tree but not as much as a giant hole. Also, dip the screws in bleach before you use them. This will prevent introducing bacteria into the tree.

STEP 5: Assembly

Now that you have everything and you know what to do, you're ready to get started. We'll begin with the anchor chain and work on up.

If you'll recall, at the beginning, I told you that the anchor should be at the bottom of the run. Now you'll find out why:

YOU DO NOT WANT TO BE ON A LADDER WHILE TIGHTENING A COME-ALONG. IT IS DANGEROUS, SCARY, AND PAINFUL WHEN YOU INEVITABLY JAM YOUR FINGERS.

Okay, sorry about that. Actually, using a come-along is super simple if you do it the right way.

1. Attach the anchor end to the anchor chain you linked around the tree. Use a link in the chain close to the turnbuckle.

2. Uncoil and attach the other end of the come-along to the loop you put into the cable. It should rest next to the jaw-end screw of the turnbuckle you attached inside the loop. You did attach the turnbuckle in the loop, right? No? Good thing you bought the jaw-end turnbuckle and not the eyebolt sort, right?

3. Tighten, tighten, tighten.

If you chose not to buy a come-along and do this using your own manly strength only to fail and sadly go back to the hardware store to buy the come-along, you will appreciate the simplicity of the tool. If you chose to buy the tool from the start, you can simply bask in the glory that you are far ahead of everyone else and basically just about done.

Once you have it as tight as it will go, attach the turnbuckle screws to the threaded buckle. If you've done this right, there should be plenty of room to tighten the buckle once you release the tension of the come-along. This is one reason you want to buy the biggest turnbuckle you can get and measure things in advance before crimping cables.

4. Release the come-along.

Please be careful. Even with the turnbuckle tightened, the come-along may still be under pressure.

5. Tighten the turnbuckle further.

Again, if everything is measured right, you'll get a wrench to help tighten the turnbuckle and there will still be room between the screw ends when you have it as tight as it will go.

Congratulations! You now have a zip line! Just put the trolley on and off you go! Now it's on to the next part: adding the camera . . .

STEP 6: Attaching the Camera, Or How I Hope My Kids Will Take Cool Footage to Post on YouTube and Earn Enough to Pay for Their Own College

To increase the geek factor of this project, I added a camera mount to the top of the trolley so my kids could connect their Flip Video camera through a standard camera mount.

Begin by creating a small platform from the remaining wood from your 1x6. Cut a 3- to 4-inch board and then drill three holes. Two holes will be for the U-bolt to attach the board to the trolley, and one hole will be for the camera mount itself.

I chose to mount the camera at the very front of the board, and I ran the U-bolt through the center of the board.

Attach the U-bolt and the board to the trolley. Bolt it down tight. For the camera mount, put the screw through the board (with a washer in between) and then attach the wing nut/washer combinations to tighten the screw to the board. Next, attach the second wing nut/washer combination to the screw, but do it in the reverse order (screw first, then washer). The idea here is that your camera will mount on the screw and then you'll use the washer and wing nut to keep it tight.

EAT, DRINK, PLAY, GEEK

Igor Bars

The Ultimate Geek Dessert

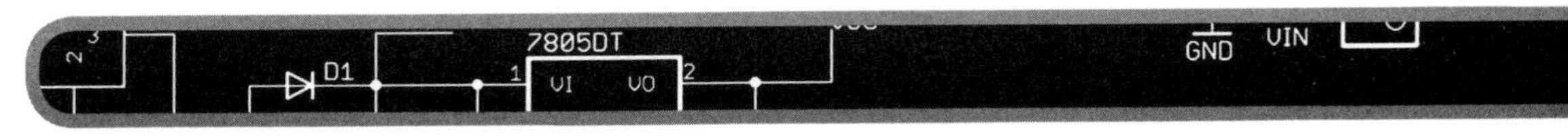

Celebrity Geek Dad Project by John Kovalic

John Kovalic is something of a legend in the Web comics and gaming communities. But even if you've never heard his name, you're probably quite aware of his work. Have you ever seen or played Apples to Apples? John created the very identifiable logo for the game, and helped develop the content as well.

But inside the world of geekdom, he's better known for the card game Munchkin (in all its myriad variations). Munchkin is a game that makes fun of role-playing games like Dungeons and Dragons, and yet has become a hugely popular title for itself as a strategy card game.

John was also one of the first comic creators to produce a regular Web comic—his Dork Tower *strip, which continues to this day to make fun of geek and pop culture using a core group of geeky characters.*

It's a testament to the power of the Internet that I've been able to connect with John. Because we move in some of the same circles on Twitter, we ended up chatting. I took a chance and invited him to co-publish Dork Tower *on www.geekdad.com, and he jumped at the offer. Since then, he's become a great friend to GeekDad, and we've done anything and everything we can to support his endeavors (mostly because they're all so geeky awesome).*

But there's one little-known side to John that this project will finally bring to light. John is a food geek as well. And he's developed what is perhaps the ultimate treat for gamers (or anyone, really), which he has agreed to share.

PROJECT	IGOR BARS: THE ULTIMATE GEEK DESSERT
CONCEPT	A recipe for an amazing and wholly unhealthy treat for gamers and geeks that you can make with your kids
COST	$–$$
DIFFICULTY	⚙ – ⚙⚙
DURATION	☼☼ – ☼☼☼
REUSABILITY	🌐
TOOLS & MATERIALS	See the ingredients listed below.

In the eight years since their creation, Igor Bars have become semi-legendary on the gaming convention circuit. I actually came up with the idea for them when I needed to find the Ultimate Excessive Gaming Snack—at least in the eyes of Igor, the most excessive character in my comic strip *Dork Tower* (www.dorktower.com). Their ensuing popularity, though, took me totally by surprise.

Since 2002, when Igor Bars were first introduced into the *Dork Tower* comic book, they've taken on a life of their own. Conventions hold Igor Bar cook-offs, and there are Web pages dedicated to them.

But two things make baking Igor Bars a particularly suitable project to share with your kids. The first is that there are three different steps that all lend themselves to children's differing kitchen abilities, so you can set kids of different ages to specific tasks that will make each of them a valuable team member. The second—and possibly most important—is that Igor Bars are almost infinitely improvisational. Igor Bar bake-offs are great fun at cons because of the huge number of variations (some far more frightening than others) that appear. Kids can have a fun time making these sweet treats THEIRS. Don't like peanuts? Add some Heath Bar pieces instead. LOVE nuts? Add some in the cookie dough, too!

The classic Igor Bar consists of a layer of pan-style chocolate-chip cookies (although any cookie that can be baked pan-style will do), a layer of peanuts and caramel (which acts as the glue that holds the bar together), and a layer of Rice Krispies treats. Cut into squares, these can serve thirty or more kids. Heck, they've been known to take down that many ADULTS. But the big trick is to keep them moist: Igor Bars should be chewy and scrumptious, not crisp and brittle.

INGREDIENTS:

PAN-STYLE COOKIE BASE LAYER

2¼ cups all-purpose flour
1 teaspoon baking soda
1 teaspoon salt
1 cup butter, softened
¾ cup granulated sugar
¾ cup packed brown sugar
1 teaspoon vanilla extract
2 eggs
1 12-ounce package semisweet chocolate chips

CARAMEL MIDDLE LAYER

2 14-ounce bags Kraft caramels
3 tablespoons milk or evaporated milk
1 teaspoon kosher salt (optional)
1½ cups dry-roasted peanuts

RICE KRISPIES TREAT LAYER

3 tablespoons butter
1 10-ounce package (about 40) marshmallows or 4 cups miniature marshmallows
6 cups Rice Krispies or other rice cereal

TOPPING

16 ounces semisweet chocolate chips

YOU WILL ALSO NEED:

15x10-inch jelly roll pan (a 9x13x2½-inch or thereabouts lasagna pan can be substituted, but thicker Igor Bars may prove tough for little mouths)
Baking parchment paper
Cooking spray, margarine, or butter to grease the pan
Two mixing bowls, one large and one small
Two saucepans
Double boiler. Two, preferably.*

Preheat an oven to 375 degrees.

STEP 1: Pan-style Cookie Base Layer

Grease the 15x10-inch jelly roll pan and line it with parchment paper. Spray cooking spray onto the paper.

Combine the flour, baking soda, and salt in a small bowl. Beat the butter, the sugars, and the vanilla in a large mixing bowl. Add eggs one at a time, beating well after each. Gradually stir the flour mixture into the butter mixture. Stir in the chocolate chips. Spread the mixture in the pan. Bake 20–25 minutes or until golden brown. (I usually go for slightly "underbaked" here, at a scant 20 minutes bake time, to keep the finished Igor Bars moist and chewy. If you prefer more solid bars, the pan cookie can be baked for a few more minutes.)

STEP 2: Caramel Middle Layer

Once the pan cookie is done, it'll need some time to cool. While it's baking is a good time to start unwrapping the caramels. This is

* Not everyone has a double boiler on hand, let alone two. You can melt the caramels and the chocolate topping in the microwave, as long as you use a microwave-safe bowl. Microwave at 30-second intervals on High, until you see things start to melt. Then continue at 20-second intervals, stirring frequently until fully melted. Be careful: The bowls will get hot.

easily the most tedious part of Igor Bars. Nobody will blame you if you designate this step to the kids' nimble little fingers. Fortunately, you can sacrifice one or two caramel squares for bribery purposes.

Place the unwrapped caramels, along with the milk and salt (salty caramel is possibly the strongest argument I can think of for the existence of a loving deity—ignore the salt if your taste and/or worldview differs) in a saucepan. Cook on medium-low heat, stirring constantly, until the caramels are completely melted. (This is a better job for the older kids. Hot caramel's nothing you want all over the floor or—more important—your progeny, and possibly you.)

You can also use a double boiler for this step, if you wish. It'll be a slower process, but there's no risk of forgetting the caramel until the pungent smell of burnt sugar tells you it's too late. If you have only one double-boiler, wash it after the caramels are poured because you'll need it later.

Pour the finished salty caramel over the cooled pan cookie and spread evenly. Sprinkle the 1½ cups of dry-roasted or other peanuts on top of this.

STEP 3: Rice Krispies Treat Layer

In a large saucepan over low heat, melt the butter and then add the marshmallows, stirring constantly until completely melted and mixed together. Remove from the heat and add the Rice Krispies®, one cup at a time, stirring them into the melted marshmallow. As soon as you have a mass of soft, crispy, chewy goodness, spread with a buttered spatula or waxed paper on top of the caramel layer.

Melt 16 ounces of semisweet chocolate chips in a double boiler. When melted, spoon or drizzle over the Igor Bars.

Wait until the bars are cooled, and then cut into 2- to 3-inch squares.

Variations

Almost anything and everything can be added to Igor Bars! Let the kids get creative. Change up the chocolate-chip cookie layer: Try a sugar cookie or oatmeal cookie base instead! Add things to the caramel layer. Switch the topping to peanut butter Rice Krispie treats. Throw on some cut-up peanut butter cups or crumbled English toffee pieces. Try a layer of frosting! Add malted milk balls! Coconut! Chopped cashews! Dare we mention bacon? Or all of the above!

Look, nobody ever accused Igor Bars of being health food. In the famous words of Cookie Monster, Igor Bars are possibly the ultimate "Sometime Food."

Measure the Speed of Light with Chocolate

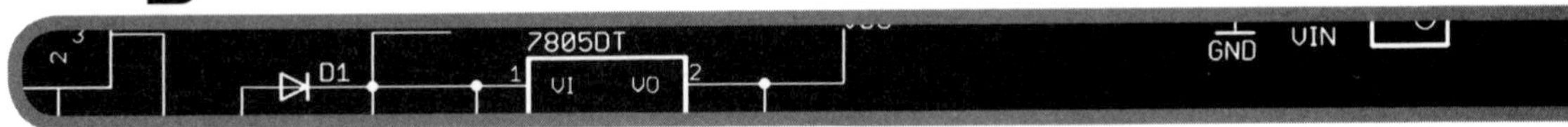

(Project Idea by Kathy Ceceri)

Food involves a lot of science (just ask supergeek Alton Brown). There are myriad chemical reactions involved in cooking and food preparation, from the crystal bonds involved in caramelization to the magic that is the emulsification of normally phobic ingredients into many tasty sauces and dressings. And while we see science going on when we cook, we rarely use cooking as a way to measure scientific properties.

PROJECT	MEASURE THE SPEED OF LIGHT WITH CHOCOLATE
CONCEPT	Cooking chocolate in a microwave will allow us to verify the speed of light with a simple measurement and some math.
COST	$-$$
DIFFICULTY	⚙ — ⚙⚙
DURATION	⏱ — ⏱⏱
REUSABILITY	🌐
TOOLS & MATERIALS	Chocolate, enough to cover an area at least 8 inches square—can be truffles, liquid-filled bonbons, or bars of solid chocolate (darker is probably better); a microwave-safe pan; a microwave; a ruler

WARNING: This experiment may take several tries to get right. I will not be held responsible for weight gained or teeth lost. To avoid familial strife, be sure to do this experiment only with your own chocolates or with candy that you have been authorized to access. You can probably find some leftover boxes on sale the weeks after Valentine's Day, Mother's Day, Easter, or Halloween.

It seems crazy, but we are indeed going to verify the speed of light by briefly heating some chocolate in a microwave. Now, there are many other materials that could work just as well, but chocolate makes a very appropriate medium, historically speaking, because the heating property of microwaves was first discovered by a scientist whose candy bar melted in his pocket when he got too close to the microwave device he was testing for use in radar. Plus, we can eat the results after the experiment is done.

The experiment works because microwave ovens produce standing waves—waves that move "up" and "down" in place instead of rolling forward like waves in the ocean. Microwave radiation falls into the radio section of the electromagnetic spectrum. Most microwave ovens produce waves with a frequency of 2,450 Mhz (megahertz, or millions of cycles per second). The oven is designed to be just the right size to cause the microwaves to reflect off the walls so that the peaks and valleys line up perfectly, creating "hot spots" (actually, lines of heat).

When we run the chocolate in the microwave for a short amount of time, the peaks and valleys of the microwaves will form hot spots on the chocolate, which will then cause localized (almost pinpoint) melting. First, we'll find the hot spots and measure the distance between them. From that information, you can determine the wavelength of the electromagnetic waves. We already know the frequency (the 2,460 Mhz mentioned earlier—though you may want to check the sticker inside your microwave to make perfectly sure), and when you multiply the wavelength by the frequency, you get the speed!

STEP 1: Bombard the Chocolate with Electromagnetic Radiation

Put the chocolate in a microwave-safe dish. If you are using separate bonbons or truffles, arrange them in a tight grid so there's as little space as possible between chocolates. If you have a bar or two of chocolate instead, lay them tightly next to each other to create a square (if possible), and if they have one printed side and one smooth side, put the smooth sides facing up.

Remove the turntable in the microwave, for we want the chocolate to stay in one place rather than move around a lot (the turntable is there to keep your food moving so the hot spots will permeate all over the food rather than being localized). You may need to put an upside-down plate over the center pillar that rotates the turntable, and then you can place your dish of chocolate on top. Close the door and heat the chocolate on high for 20 seconds.

STEP 2: Inspect the Irradiated Candy

Open the door and, using a flashlight to increase glare, look for hot spots. Depending on the candy you use, you may have to feel it to see where it has softened. With liquid-filled cordials, you may see several shiny spots and even spots where the chocolate shell melted through and released the sweet syrup inside. Chocolate bars may just show a series of small shiny dots. If you see none of these, close the door and run for another 10 seconds. Check and repeat if needed until you see the spots. Then take the pan out of the microwave.

STEP 3: Measure and Calculate

Using your ruler, measure the distance between two adjacent spots (geeks like to use metric, so that's what we'll do here). The space between two spots should be the distance between the peak and valley (crest and trough) of the wave. Since the wavelength is the distance between two crests, multiply by 2. Finally, multiply that result by the frequency expressed in hertz, or 2,450,000,000, or 2.45×10^9 for those learning scientific notation. Remember: If you checked the sticker in your microwave and it showed a different frequency, use that instead).

In our trial, we measured a distance of roughly 6 centimeters: 6 x 2 x 2,450,000,000 = 29,400,000,000 centimeters per second, or 294,000,000 meters per second. This is awfully close (less than 2 percent difference) to 299,792,458 meters per second, which is the speed of light. Not a bad error quotient for some leftover chocolate and a kitchen appliance!

If you want to use some proper scientific protocol, you could perform a series of tests and do some simple statistics to refine the answer by establishing your standard deviation and determining your results within a specific degree of certainty (perhaps this higher-level math would be good for your teenage geeklings?). This would require you to eat quite a bit of chocolate so that it doesn't go to waste after each test. Not that I would support that kind of thing. But it IS in the name of science!

Toy Candy Molds

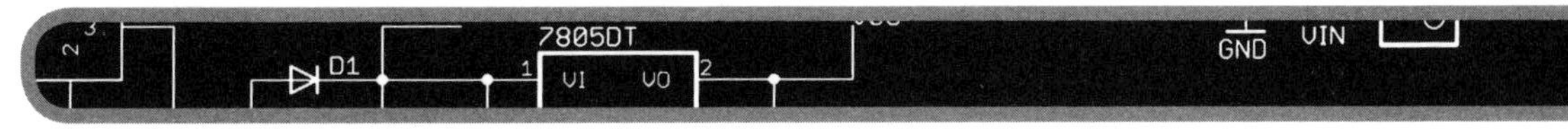

(Project Idea by Natania Barron)

Here's a project that is perfect for a geeky birthday party or even just a special treat for the family on a weekend afternoon. Because any project that involves candy making is a surefire way to keep kids interested and excited!

The magic of candy making is a great example of how science works. But to make it even geekier, instead of making chocolates or candies in shapes we're used to, let's take candy making a step further. How about we make chocolate superheroes? Or Jell-O race cars? Or marzipan LEGO minifigures?

Yeah, that sounds more like it!

PROJECT	TOY CANDY MOLDS
CONCEPT	Use food-grade silicon molding compound to quickly and easily create custom molds out of your kids' favorite toys.
COST	$$–$$$
DIFFICULTY	⚙ — ⚙⚙
DURATION	⏱⏱⏱ — ⏱⏱⏱⏱
REUSABILITY	🌐🌐 — 🌐🌐🌐
TOOLS & MATERIALS	Food-grade silicone molding compound (Silicone Plastique or other); chocolate, or other moldable filling; toy of choice

The main idea of this project is pretty simple: Use specially made food-grade mold compound and a few toys chosen by your kids to create molds. You can find silicon molds on www.amazon.com, www.makeyourownmolds.com, or in your local crafts or cooking store. Then you'll fill those molds with some kind of candy or confection that starts in a liquid state but then cools to a solid, allowing you to extract a pristine replica of the original toy, made of the tasty treat. The devil is, however, in the details, so please read on:

STEP 1: Choose, Clean, and Dry Child's Toy

The best toys are the simple ones, like toy cars or LEGO blocks. You really don't want a lot of things sticking out at odd angles, or a lot of interior space for the liquid mold to seep into. So, for example, a Hot Wheels car with windows that are always closed works well. A slightly larger model car that has its windows down and a detailed interior would be bad. However, a lot of surface detail is okay and will be re-created in the mold rather amazingly.

For larger or more detailed toys, consider doing a mold of each half and then joining them together at the end of the project. If you use something like chocolate as your candy material, it's pretty simple to warm the joining surface after it's out of the mold and just stick the two halves together, then put them in the freezer to chill and set.

STEP 2: Combine Molding Compound per Instructions

This varies by brand but usually requires mixing two separate compounds into one. The best we like for this kind of project is the sort that starts as two claylike compounds of different colors (e.g., Silicone Plastique). They get kneaded together in equal portions until they are blended and the material is ready to make the mold. It's a similar material to clay or Play-Doh, which makes it very familiar for smaller geeks.

STEP 3: Make the Mold

Estimating how much compound you'll need for each toy will take a little eyeballing, but try to get enough so that it could envelop the toy (even though we won't quite do that). Gently press the compound around the toy so that it blankets nearly the entire toy, but be sure to leave enough of an opening in which to pour chocolate or other filling.

The best strategy is to try and keep one side flat, which will make the setting process easier later on (otherwise the molds will roll around). If the toy has a "top" and "bottom," press the top of the toy into the center of your blob of compound, and then work the compound around the sides of the toy and even over the bottom a little, but make sure you leave enough room to get the toy back out—if you want to reuse the molds, this is particularly important.

Then rest the compound-covered toy on your work surface, with the top facing down, and press gently so the compound squishes a little to create a flat surface, enabling the whole thing to lie flat on the table, which will make it that much easier to let it sit and cure, and later to let your candy fillings set and harden inside the mold.

Let the compound cure per the mix's instructions. Typically this takes about 20–30 minutes. This could be a good time to start prepping your candy fillings!

STEP 4: Remove the Toy and Clean the Mold

When the mold has set, remove the toy from inside it. Silicon has some pretty good stretch but sometimes has a habit of tearing, so it may be best for parents to do this part.

First try pulling the mold away at the edges a bit. Work around the perimeter of the mold, tugging lightly every little bit until it's clear the mold is loose of the toy. Then you should be able to pop the toy out like turning a sock inside out.

Gently clean the molds to make sure nothing has stuck to the inside. You may need to trim off small burrs made by odd shapes in the toy. You can either use small scissors or just your fingers. Now you have a perfect mold of the toy!

Before putting food materials in the mold, wash the mold with mild soap and water (first make sure the compound instructions allow this).

STEP 5: **Prepare Ingredients for Filling**

Chocolate is the easiest (and perhaps most delicious!) filling to try first, for it melts easily and sets up very hard. Select your chocolate of choice—white, milk, or the various levels of dark (our favorite)—and melt the chocolate in a double boiler (with water in the bottom) over medium-high heat.

Using a double boiler is key because it melts the chocolate in a much gentler, more gradual way, making sure it doesn't burn (as it might if it were in a normal pan sitting directly on the stove heat). If you're unsure how much chocolate to melt, try the buoyancy test (ah, SCIENCE!): Drop the toys into water in a large glass measuring cup and see how many ounces the water level fluctuates—that's how much chocolate to melt.

Be careful not to let the chocolate get too hot, for the cocoa butter will separate. And if for some reason you think it is getting too hot, don't put water into it to cool it down. You'll get an unusable sludge. Just take it off the heat for a while.

STEP 6: **Make the Candy**

Now it's time to pour the melted chocolate into the molds. For smaller molds, pouring straight from the pan or from a creamer or measuring cup with a pouring lip is ill advised. Put the melted chocolate into a durable ziplock bag and snip one of the bottom corners to make it serve like a pastry bag. This allows for far more precision.

When your molds are filled, shake them to ensure there are no air bubbles. Obviously, don't jar them so much that you spill chocolate (at which point the candy police will show up and issue you a citation), but put enough vigor into it to do the job.

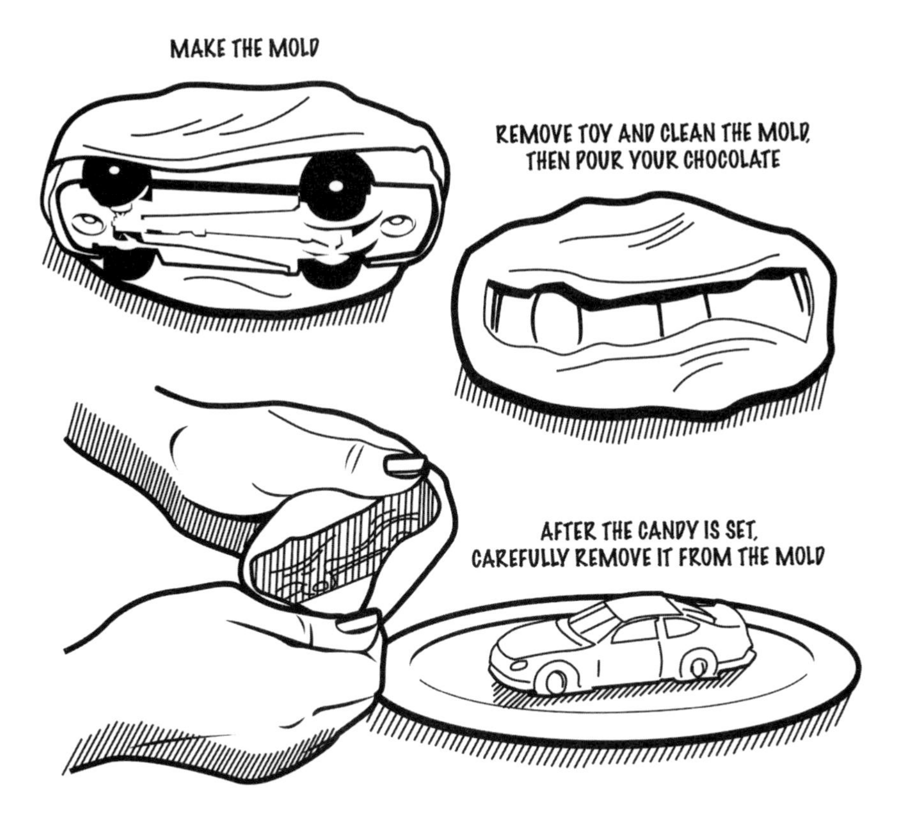
MAKE THE MOLD
REMOVE TOY AND CLEAN THE MOLD, THEN POUR YOUR CHOCOLATE
AFTER THE CANDY IS SET, CAREFULLY REMOVE IT FROM THE MOLD

Now the hard part: You need to refrigerate them overnight. This is going to be the best way to make sure they set all the way through. You could try the freezer, but you're more likely to get a really hard shell of chocolate and have to wait to eat them anyway. If you're feeling a bit more scientific, you could try using some of the dry ice from the Dry Ice Ice Cream project elsewhere in this book as a supercooler. Or spray them with compressed air to speed up cooling. It's up to you.

When they're set, remove the chocolate from the mold in much the same way you removed the toy when you made the mold originally. You should now have a carbon (or chocolate) copy of the toy. You may need to shave a little off here and there, since the chocolate will change shape slightly during cooling, but the end result is quite impressive.

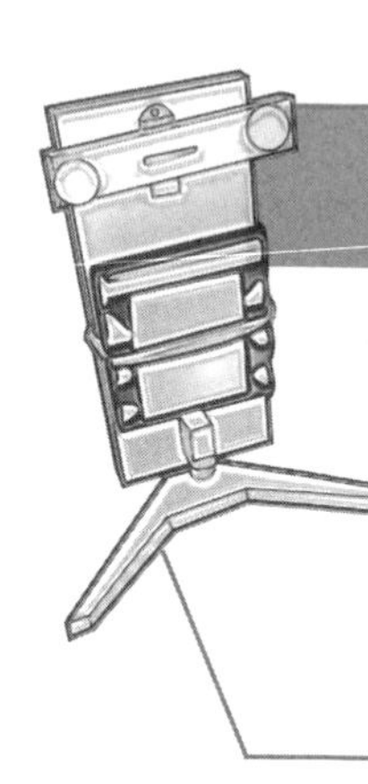

Extra Geeky Ideas

Chocolate should just be your start! Indeed, if you want something that is faster to prepare than chocolate, try Jell-O or similar gelatin desserts, since they'll set far more quickly.

Depending on the mold material you use, you may be able to make hard candy in your molds as well, but check first. Most hard-candy recipes require the molds to handle 300-degree temperatures or above.

Dry Ice Ice Cream

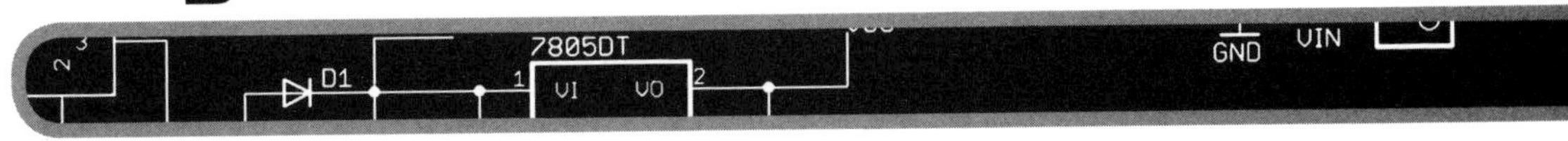

(Project Idea by Kathy Ceceri)

Carbon dioxide is a somewhat magical material (even if everything magical about it can be explained . . . BY SCIENCE!). It is one of the few materials that at normal atmospheric pressures (meaning what we live at) will subsume—that is, transform from a solid to a gas without bothering with that pesky liquid state we're so used to in other materials. Regular water (H_2O) ice, when it is heated, will become liquid water first, and then start steaming away (or evaporating, depending on the temperature). Heck, iron melts first. I'd hate to see what it takes to get iron gas.

But carbon dioxide, if set outside as a frozen block, will just slowly vanish in a white fog. It's more exciting if you put it in a pail of water—it bubbles violently, and the bubbles are filled with white gas. Even better, it freezes at -78 degrees Celsius, meaning it's not that hard to make (no, we're not going to make it, but it's not that hard or expensive to get).

So, in a way, it's like super-ice. It's colder than regular water ice, so it'll keep whatever you want cold colder than water ice, and when it melts, it doesn't leave a puddle. So what can we do with it?

Well, on a hot summer day, I can remember one very important thing we used to use good old normal water ice for: making homemade ice cream. In fact, we had to add salt to the ice to make it colder to aid in the chilling process. Why don't we do away with all that fuss and muss, and use a much faster method instead?

PROJECT	DRY ICE ICE CREAM
CONCEPT	Use dry ice (frozen CO_2) rather than water ice to chill homemade ice cream.
COST	$–$$
DIFFICULTY	⚙⚙ – ⚙⚙⚙
DURATION	☼☼ – ☼☼☼
REUSABILITY	🌐
TOOLS & MATERIALS	Dry ice; heavy cream; milk; sugar; vanilla

Obviously, the elephant in the room on this project (metaphorically speaking—we're saving the exotic animal projects until the third book) is the dry ice itself. The first area of concern is safety. However, dry ice is safe to use with kids, as long as you avoid touching it with your bare skin or getting it in your eyes. Try wearing winter gloves and using tongs to handle the ice or anything going near it; safety goggles are also a must and always add a scientific flair to any activity.

The second area of concern is actually procuring the dry ice. You don't just run down to the supermarket to pick it up! Indeed, you may be afraid it's going to be too hard to find. However, you may be surprised.

If you live in any reasonably suburban-to-urban environment, it probably won't be that hard to find some dry ice. Do a quick Google search (or, um, use your Yellow Pages) and look for ice companies in your area. There are often more than one in a metropolitan area, serving a wide variety of industrial and food-service customers. Give them a call and see if they'll sell you some dry ice. There may be a minimum amount—say a ten-pound block—but whatever you don't use for ice cream can be used for other fun. Make sure you ask them if they'll supply a cooler, or if you need to bring your own. And make sure you do this the day you plan to make the ice cream. Dry ice will not stay frozen overnight in your freezer. The recipe itself is basically the same as what you'd do with an old-fashioned (or even ultramodern) ice cream maker.

DRY ICE SINGLE-SERVE ICE CREAM RECIPE

½ cup heavy cream
¼ cup milk
1 tablespoon sugar
¼ teaspoon vanilla
5–10 pounds dry ice in a block (depends on what your local supplier sells)

1. Put all the ingredients (except the dry ice) into a small ziplock plastic bag and seal it shut, pressing out as much air as possible.

2. Shake it all around to mix the ingredients.

3. Lay the bag on the slab of dry ice for 10 seconds (count them out loud, using the "Mississippi" method for accuracy). The dry ice can be in your sink (where you'll get a lovely pool of mist) or on any ice-safe surface outside.

4. Pick up the bag carefully (again, use the tongs for safety) and squish it about with your gloved hands, shaking it, deforming it (don't break it), and re-forming it for a few seconds.

5. Repeat Steps 3 and 4 until the bag achieves the firmness you like (soft-serve or classic box-serve).

The basic idea is that the dry ice is so much colder than water ice that it'll chill the ice cream base a lot faster than a normal ice maker. Using smaller bags and batches makes it go even faster and can be the perfect means by which kids in a group each gets his/her own done quickly on a warm day.

For the adults, you can try using a larger bag and multiplying the recipe for bigger batches. You can also try finding high-quality ingredients (organic cane sugar, bourbon vanilla) to make the result tastier, and/or you can experiment with recipes. Any ingredient you can find that works for a traditional ice cream maker will work here as well.

To double-down on the geekiness of this project, try a special ingredient GeekDad.com Assistant Editor Matt Blum suggests. Take a couple strips of bacon per serving of ice cream, coat them with brown sugar, and cook them in your oven at 325 degrees for about 30 minutes (but keep an eye on them. Best to do them on racks in a baking pan so the fat drips off and they bake evenly on all sides). Then break them up into small pieces (you can also chop them with a butcher knife), and include them with the rest of your ingredients in your plastic bag for freezing.

More Geeky Fun

CO_2 freezes at about -78 degrees C (-109 degrees F), so it's handy for making things super cold, super fast. It also creates interesting effects when combined with everyday substances like water. If you've never used dry ice at home before, finding a few quick and easy other activities for your leftovers takes only a little searching on the Internet. Here are some ideas that we've tried in my family:

- Drop a chunk of dry ice in a plastic cup of water and watch it boil away merrily while "steam" pours from the top. This is the simplest, classic Halloween cauldron setup, only on a miniature scale.
- Try adding a few drops of dish soap and get instant overflowing bubbles.
- If you slowly draw a paper towel soaked in soapy water across the rim of a bowl filled with water and dry ice, the off-gassing CO_2 produces one giant bubble.
- Put chips of dry ice into small ziplock bags and seal them. Watch them "inflate" as the ice subsumes, and eventually pop with satisfying effect.
- Make a "Hero's Engine" (Hero was a first-century A.D. Roman mathematician who described the concept of using escaping gas as a means to create mechanical energy). Take a plastic milk jug and tie a string to its handle. Hold the jug by the string to find its equilibrium and then poke holes in the sides on each concentric corner near the bottom. Drop a few chips of dry ice into the jug, put the cap on, and hold it by the string. As the dry ice subsumes and the CO_2 expands, it'll create little jets out of the holes and start spinning the jug.

Homemade Root Beer

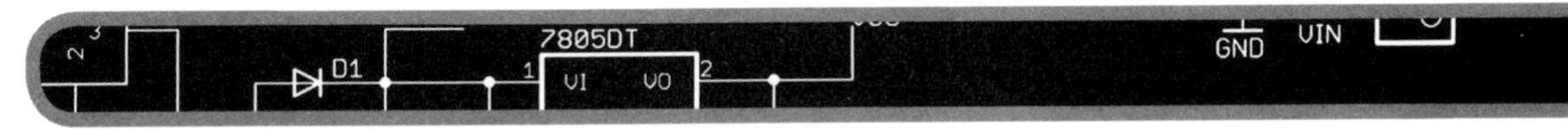

(Project Idea by Kathy Ceceri)

Root beer. Snoopy's favorite drink when he is taking time off between battles with the Red Baron. It's a soda that just seems that little bit more wholesome than the big-brand colas because of its old-school cred. Indeed, now there is a whole submarket of "craft" root beers that try to replicate the original process of making it from the treated root of the sassafras tree (from whence also comes sarsaparilla, but that's another chapter in another book).

But what is root beer really? Like most other sodas, it's pretty simple: Root beer is a sweetened, flavored water with carbonation added. Seems like a pretty easy formula to make yourself, right? Right!

The sweetened, flavored water is pretty easy to do using available flavorings (less so if you actually want to cook down some sassafras root). It's the carbonation that can be challenging. Of course, we don't have industrial machines to do the job, so what's the next best way?

Here's a hint: It's why *beer* is part of *root beer.*

Yup, we're going to ferment the root beer, just a tiny little bit. There will be enough alcohol in the mix when you're done that if you drank, say, ten gallons, you might feel a buzz coming on before you rolled into the bathroom. So, in other words, it's negligible.

The best part is that it's pretty darn easy to make at home and is an excellent primer of the science of yeast and fermentation. So, while you may just about break even on price with this project, the education it imparts (and the, hopefully, tasty drink at the end of the line) is priceless. Okay, maybe not priceless, but it'll pay you back double!

PROJECT	HOMEMADE ROOT BEER
CONCEPT	Brew your own root beer at home.
COST	$–$$
DIFFICULTY	⚙ – ⚙⚙
DURATION	☼☼☼☼
REUSABILITY	🌐 (but easy to repeat)
TOOLS & MATERIALS	A clean 2-liter plastic soft drink bottle with cap; funnel; measuring cup; measuring spoons; 1 cup table sugar; 1 tablespoon root beer extract; ¼ teaspoon baking or champagne yeast; water (preferably distilled)

Home brewing beer has been a hobby of mine since, well, since I had a garage to do it in. Back in the day, my buddy Randall and I brewed a batch we dubbed Lightsaber Lager, and even printed up labels using the classic *Star Wars* poster with Luke and Leia on it as the foundation base image (I won't mention here what we did to it in Photoshop to . . . enhance certain features).

[DANGER: THE FOLLOWING PARAGRAPH INCLUDES SOMEWHAT GROSS ANALOGOUS DESCRIPTIONS OF BIOLOGICAL PROCESSES TO EXPLAIN BREWING BEER. YOU HAVE BEEN WARNED.]

So I love home brewing. Which, as a hobby/geek passion, contains equal parts science and magic (that is, cooking, which may be based on science but quite often seems magical as well). When you brew beer, you make a heavily sugared liquid base using water, hops (a bitter herb), and malt (concentrated sugars made from caramelized grains), and then you add yeast and let the solution ferment for a couple weeks. During that time, the yeast eats the sugars and, well, poops the alcohol and farts carbon dioxide. At the end of the process, you filter and bottle the result, and add a bit more sugar before capping. The yeast eats the residual sugar and, in the pressurized environment, farts a bit more, which carbonizes the beer.

[SORRY ABOUT THAT.]

Now, we're not going to brew beer here. We're going to brew root beer. So we're going to bypass that whole "couple weeks" of fermentation and skip straight to the part where we have a finished product that we bottle and carbonize with the little bit of extra sugar.

Obviously, the easier way to make soda is just to experiment with mixing flavorings with unflavored seltzer, but the method we'll do here is more scientific and has the added fun benefit of being risky: If you let the yeast ferment too long, the pressure of the carbon dioxide inside the bottle may cause it to explode. That's why it's recommended to do your own home brewing in recycled plastic bottles. And leave them in the garage. Covered.

Root beer flavoring is rarely made from scratch (except in the case of trendy, craft beers) and there are hundreds of different flavorings to choose from. You can find many brands online (search for "root beer flavoring" on Amazon to see an amazing array). But you can also use just the standard root beer concentrate from the spice aisle in your supermarket with good results.

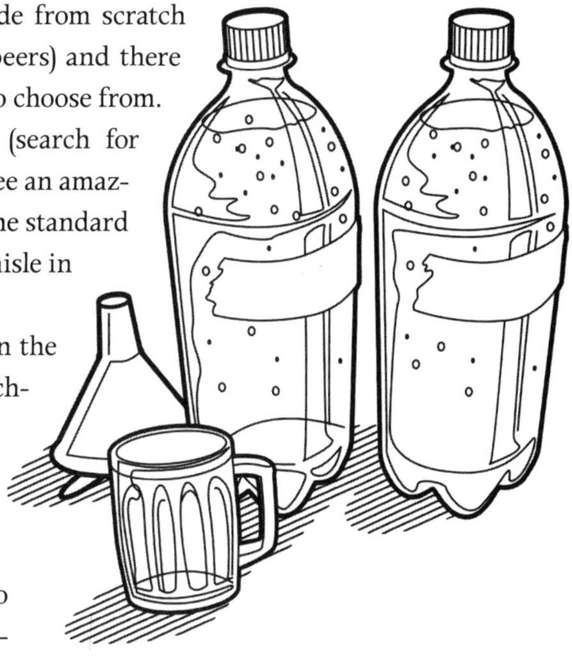

Yeast is sold in strips of packets in the baking aisle of the supermarket (Fleischmann's is a famous brand often used for baking, but you can use any brand for this project). Each packet holds a little more yeast than you'll need for a 2-liter bottle, so be sure to measure it or plan for multiple bottles. You can also buy it in bulk; it keeps very well in the freezer. Some root beer makers prefer champagne yeast, which can be found in home-brewing stores (yet another reason to get acquainted with the overall hobby of home brewing). Indeed, you could always do this project in sync with your own beer brewing, to include your kids in the science and fun of the practice.

STEP 1: Preparation

One key part of the whole operation is having clean parts and tools. Wash everything with very hot water or, even better, soak them in boiling water. Foreign agents (bacteria, not spies) can kill the carbonation process.

STEP 2: Mixing

Use the funnel to pour the sugar into the clean, dry bottle. Add the yeast and shake it up to thoroughly mix it with the sugar. Pour in the root beer flavoring. Fill the bottle halfway with water. Swirl to get any ingredients that are stuck to the side of the bottle into the mixture. Pretend you're Tom Cruise in *Cocktail*. Or Brian Brown. Probably Brian Brown.

STEP 3: Bottling

Fill the bottle up to the neck with water, leaving about an inch of air space. Shake to mix completely. Put the cap on tightly and use a marker to mark where the water line is at the time of capping. Then let the bottle sit at room temperature for about three or four days—less if it's warm. This is where I suggest you leave the bottle in a nice out-of-the-way spot in your garage so that if it explodes, not much will be damaged (either by the explosion itself or by the sugar water that would then be all over the place).

STEP 4: Serving

Once the bottle is hard when you squeeze it, and the water level has gone down two inches (measure this against the mark you made at bottling), refrigerate. Be careful not to let it sit out past the point where the bottle is too hard to squeeze. Refrigerate the bottle overnight. To open, untwist the cap just a little to let out some pressure. Then open the cap the rest of the way.

Serve in tall frosty mugs. Some yeast sediment may remain at the bottom of the bottle, so be careful not to pour the last little bit out.

That's it. You've made homemade root beer. You may want to play with the recipe to get something that's just right for your taste, but you've got the methodology down. The rest is mixology, which is easy.

And you also might well be tempted to try this carbonation to make other flavors of soda. "Cheers!" we say. However, you'll want to keep one thing in mind: The fermentation process does add a certain yeasty flavor to the results, so using just any fruit juice probably won't lead to a satisfying result. You're better working flavors based on spices or other roots, like ginger beer, birch beer, or cream soda. But again, experiment!

Oh, and considering there's a project on how to make a very geeky sort of ice cream elsewhere in this book, you also now have the tools to make the geekiest root beer float ever. I trust you'll use this knowledge for good.

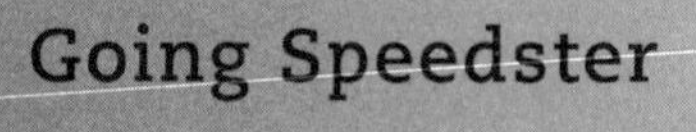

Going Speedster

A Celebrity Geek Dad Anecdote by Rod Roddenberry

Eugene Wesley "Rod" Roddenberry, Jr,. is a board member of the X PRIZE Foundation, avid diver, movie and television producer, and all-around cool guy. He's also the son of one of the most famous geeky couples ever: Gene Roddenberry, creator of Star Trek, *and Majel Barrett-Roddenberry, who played Nurse Chapel in TOS (The Original Series—how proper fans refer to the original television show) and Lwaxana Troi on* TNG *(*Star Trek: The Next Generation*). When I asked him about including something in this book about being the son of a renowned Geek Dad, he very graciously sent me this wonderful story about the great days he and his dad would sometimes have together, as an idea for how all geeky parents can spend fun time with their kids.*

When I was a kid, my dad and I would do something we called "bumming." Basically, it was just a day out together, usually to the Santa Monica Pier, but he always framed it in this cool way that made it something more special.

First, I'd get to sit in the front seat of the car with him, which is always cool for a younger kid (and probably illegal or inadvisable these days—as so much fun stuff is), and I could barely see above the dashboard. We would head out from the house, and when we got on the freeway, he would turn to me and say, "Okay, Rod, now we're going to go Speedster!" This was the magic word that meant we were going to go super fast, like warp speed.

Suddenly, the car engine would rev up really loud, and we'd be flying down the freeway! Looking back, I know all he did was to downshift the car to increase the engine RPMs, but at that age, it sounded like we were trying to break the sound barrier!

We would get to the pier and he would give me quarters for the arcade or buy me the balls to knock over the milk jugs and win a prize. Or he'd win one for me. We'd have lunch somewhere along the pier or somewhere in Santa Monica, and just generally have a great day out together.

Then we'd head home. On the way, we might do some speedstering, and then when we made it home, we'd pull into the garage and close the door, but he would leave the car lights on. Then he would lean over to me and say, "Don't tell your mother we did this!" in a conspiratorial way (even though she probably knew quite well what we'd been out doing).

We would end it with him saying, "I am Gene 'Poo-Poo' Roddenberry," and I would respond, "I am Rod 'Poo-Poo' Roddenberry." That was our ceremonious way of ending the day.

Of course it was just a regular day out, but it was special because it had this ritual to it—this insider's language and ceremony that made it all our own, our secret. That's the kind of thing that can turn just a day into something magical that you remember forever.

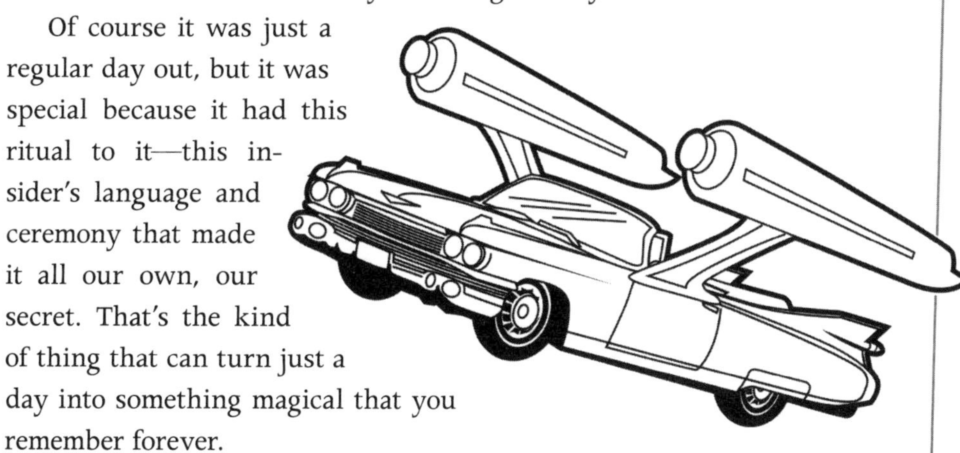

CRAFTY LIKE A GEEK

Clothes Hacking

Mod Your Kids' Clothes with Stencils

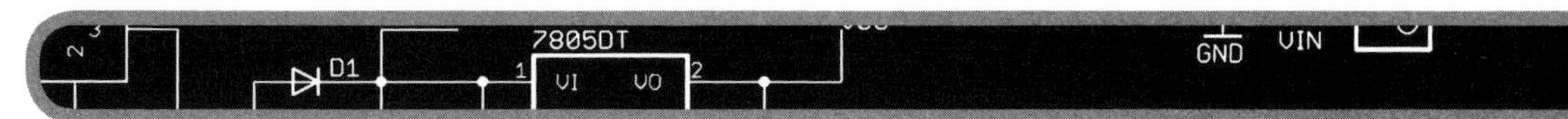

A Celebrity Geek Dad Project by Patrick Norton and Sarah Holm Norton

Patrick Norton is a tech journalist well known to the computer geek community. He was cohost of the groundbreaking Screen Savers *on TechTV, and later of the gadget video podcast* DL.TV. *He is currently a managing editor at the Revision3 network, where he hosts the shows* Tekzilla *and* HD Nation. *Sarah Holm Norton is a records librarian who has worked on document management projects for NASA and the EPA. She started crafting after becoming the lead developer of their son, Seamus.*

Is there any item of clothing more Geek than the T-shirt? Sadly, while Mom's sporting the I VOID WARRANTIES tee, and Dad's wearing his favorite anime character, all too often the racks at the local children's store are packed with irritating cartoon characters from a certain Gigantic Corp That Changed Copyright.

Enter the punk rock art of stenciling, which works for clothing just as well as it does for sidewalks and walls.

Stenciling your kids' clothes or, better yet, helping your kids stencil their own clothes, allows you to create a charming, unique wardrobe without blowing a fortune. Bonus: Giving kids control over their clothing can help resolve dressing strikes staged by young nudists. And best of all, the skills that produce a pair of rocket-stenciled underpants or monkey-faced T-shirt today will easily produce custom tees for the rest of your child's life. (If she produces rude T-shirts in high school, we're not responsible!)

PROJECT	CLOTHES HACKING: MOD YOUR KIDS' CLOTHES WITH STENCILS
CONCEPT	Create customized T-shirts (and underpants!) using stencils and paint.
COST	$–$$
DIFFICULTY	⚙
DURATION	☼☼
REUSABILITY	🌐🌐🌐🌐
TOOLS & MATERIALS	Prewashed cotton T-shirts; permanent fabric paint; 1-inch sponge brushes; masking tape (cheap brown works just fine; save your fancy blue or green autobody masking tape for painting stripes!); sturdy piece of cardboard or other flat surface to use as backing; small dishes or a cupcake pan; an iron and a cloth for pressing; newspaper or other table protection is recommended; a drying rack or place to let your T-shirts dry is recommended

We use premade stencils from a variety of sources, but you should be able to find some at your local arts and crafts supply store. Or search "stencils" on Amazon. Ed Roth's *Stencil 101: Make Your Mark with 25 Reusable Stencils and Step-by-Step Instructions* is especially awesome, and costs under $20.

STEP 1: Prep for permanent paint mayhem! You can clean up most permanent fabric paint with cold water if you move quickly. (Check the label before you spill or spatter!) If your kitchen table isn't ruined already, we suggest you cover it with newspaper or craft paper. Then get everybody decked out in paint-friendly clothes, a smock, or both. Ready?

STEP 2: Lay out your clean T-shirt to figure out where you want to place your stencil, and to make sure there are no wrinkles. Take your backing and insert it into the shirt, arranging the fabric so it is taut but not stretched out. (Horizontal wrinkles across the chest of the shirt are a sign that your backing is too wide for the T-shirt!)

STEP 3: Attach the stencil to your T-shirt with the masking tape. A bit of tape on the corners is usually enough. You want the stencil to be flush against the fabric to keep the shape edges sharp.

STEP 4: Plop some paint into your dishes/cupcake tin. We use an old supermarket-bought cupcake pan as a quick and dirty palette that keeps paint colors separated. Start with 1–2 tablespoons of each color (you can always add more!), one color per dish/cupcake mold, and one dry sponge brush per color. Stenciling with toddlers guarantees some paint mixing, but having all of those brushes often buys us five minutes of chaos-free fabric modding.

STEP 5: Start painting! Before you grab your paintbrush, press the edge of the stencil flush to the fabric to keep your image crisp. Dip your dry sponge brush into your first color and dab paint onto your T-shirt, starting along those pressed edges and then moving toward the middle of the design. Press, paint, then move your fingers to the next part to press down for painting!

You'll want to use enough paint to fill in your image/phrase, but not so much that you soak through your shirt, because the paint will stiffen the fabric.

STEP 6: Remove the backing and allow the paint to dry thoroughly. This can take 12–24 hours, but always check those paint labels to be sure. A drying rack is helpful, as is a clothesline, a hanger, or a table where the paint can dry undisturbed. Err toward drying the shirts for longer than you think they need to dry, especially if you're in a humid part of the world!

INSERT BACKING INTO SHIRT
ATTACH THE STENCIL TO YOUR T-SHIRT WITH MASKING TAPE
START PAINTING
FINISHED

STEP 7: When the paint is completely dry, it's time to heat-set the paint. Use an empty steam iron and a pressing cloth (a tea towel, flour-sack towel, or a pillowcase will all work; just don't use a looped fabric towel!). Set the iron on "High" or "Cotton," cover your design with the pressing cloth—no wrinkles here—and apply steady heat to all of the design for 2–3 minutes.

STEP 8: Wash and dry your stenciled clothes before you wear them the first time. It's a good idea to wash them alone. Nobody wants their paint to run over the rest of the wash!

STEP 9: Wear 'em till they (or you) outgrow 'em!

Ready to move beyond store-bought stencils? Freezer-paper stenciling is an increasingly popular way to customize clothing. That's because they guarantee you clean lines for your design. The downside is that each stencil can be used only once because you literally iron it onto your fabric.

You'll be using the same painting and setting techniques listed above with two substitutions. First, you'll need to make the stencils. Trace your (or your kids') favorite images onto freezer paper, then cut them out with a hobby knife.

Unless you have older children capable of cutting out their own stencils, you'll probably want to make these stencils ahead of time.

Second, instead of grabbing the masking tape, you'll heat up that dry iron, insert your backing into the shirt, and align the freezer-paper stencil on your shirt. Next, iron your custom stencil to your shirt with the shiny side down. Go ahead and paint on your image, and once it's dry, you can peel off the paper and set the image per the instructions above!

Need some ideas for stencils? GEEK 2.0 is a classic for children's T-shirts, though you'll be making this one yourself. . . . We're contemplating ADVENTURE IS OUT THERE for our next road trip.

Freezer-paper stencils are a good way to replicate simpler cartoon characters for your child. Just remember, the more complicated the character, the more time you'll be spending with a hobby knife, cutting tiny little bits of paper. . . . Starting simple is a very good idea here!

The monkey face, bluebird, hummingbird, and spray paint can from Ed Roth's *Stencil 101* are very popular with our toddler. You can also buy smaller, 4x2¾-inch and 4x6-inch one-sheet stencil collections directly from the Roth Web site, www.stencil1.com, or, if you're looking for something a bit bigger, 8½x10-inch stencils.

Make Your Own Mini-Me

A Step-by-Step Guide to Customizing LEGO Minifigs

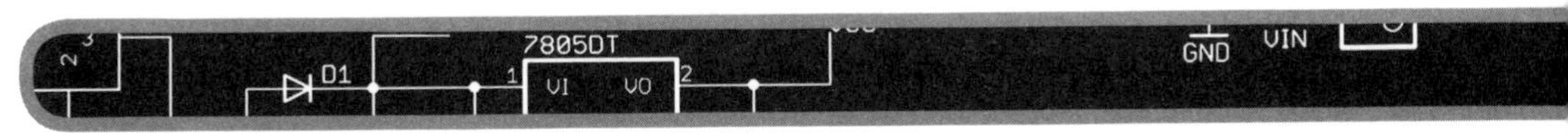

(Project Idea by Dave Banks)

LEGO minifigures—the little people that come as action figures in many LEGO kits—are wonderful fodder for the imagination. They are perhaps the most basic toy-representation of a human figure on the market, and yet they are so fun and whimsical that children can engage in extensive imaginative play with them.

But the variety of minifigs available is limited. There are a number of basic designs, and then there are the special versions that come with certain kits (the *Toy Story* and *Star Wars* sets are really cool). But if you want to do anything original, well, your options are pretty limited.

Until now.

PROJECT	MAKE YOUR OWN MINI-ME: A STEP-BY-STEP GUIDE TO CUSTOMIZING LEGO MINIFIGS
CONCEPT	Use printable decals to dress up your LEGO minifigs or even add your face to them.
COST	$$–$$$
DIFFICULTY	⚙⚙ — ⚙⚙⚙
DURATION	⏱⏱⏱ — ⏱⏱⏱⏱
REUSABILITY	🌐🌐🌐 — 🌐🌐🌐🌐
TOOLS & MATERIALS	Bottle of Brasso (or similar cleaning agent); cotton swabs; old towels or rags; scissors; ruler; hobby knife; tweezers; paintbrush; decal setting solution; distilled water; clear gloss spray paint; laser or inkjet clear decal paper for 6 sheets; LEGO minifigs (head, legs, and torso can be bought in a variety of colors at http://shop.lego.com/PAB/ [under minifigures]); home computer; illustration software; image editing software; inkjet printer

Perhaps you'd like to create a vast army resplendent in mauve and high-gloss black uniforms, which you've chosen to represent your house's coat of arms. Or you could even make minifigs based on the cast of *Cats*, in all their plastimagorical wonder. Or, well, how about personalizing your minifigs and creating versions of you and your geeky family? This project will give you the skills to achieve any (or all) of those ideas here, by showing you how to alter the markings on the ordinary generic minifigs you can purchase in bulk from LEGO (or just fish them out of the box on the shelf where all your bricks are stored).

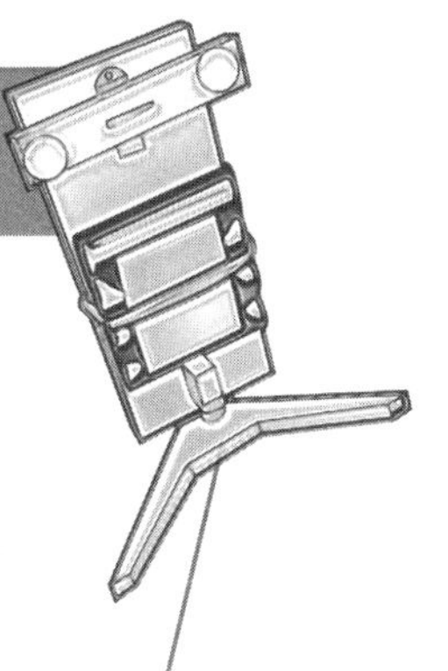

Geek Dad Book Flashback Moment

In the first Geek Dad book, I included a project on how to make your own cartoon strips, using action figures—including self-altered LEGO minifigs—as your characters. In that project, I told you about simply scraping the features off with a knife and then drawing new features on with a felt-tip pen. This project takes that idea to a whole new level.

STEP 1: Creating the Decals

Yes, we're using decals. You probably haven't played with decals since you were a kid, and here's your chance to pass on that experience to your kids at just the same age. When they first ask, "What's a decal?" you just tell them: "It's like a sticker, but with a transparency layer." Oooh, digital graphics humor.

The cool thing is that these days, not unlike homemade cards, iron-on transfers, and labeling stickers, there's inkjet printer stock for making decals as well. You can get 10 sheets for $1 to $2 per sheet (so yeah, it's not the cheapest thing in the world, but because we're skinning minifigs, we can make it go a long way) depending on where you find it online. A quick Googling of "inkjet decal paper" will bring up a wide selection, but I suggest you look for the brands being sold by model train enthusiast Web sites, as those folks are geeks as well, and they are obviously passionate about the materials they use for their models.

Since you are doing this from scratch, you'll have to make your own templates for the decals that you'll add to different parts of your minifigs. I've made some templates available on www.geekdadbook.com, but I promise you, this isn't going to be that tough. You might need to do a little trial and error by printing out your drafts on regular paper, cutting them out, and fitting them to your minifigs as test samples, but that's pretty easy.

The decals we're making in this project are for the face of your minifig, and for the torso (so the clothes). You're going to start in some kind of drawing program on your computer, preferably one that makes drawing shapes easy (a fun choice, especially for younger kids, is Tux Paint, which is totally free and available for Macs, PCs, and Linux machines: www.tuxpaint.org).

Let's start with the torso. The torso should be a trapezoid 0.6 in. at the widest point and about 0.5 in. high. If your drawing program can draw trapezoids, start with that—either as an outline for your decal or simply as a guide for whatever designs you're going to use. The rest is really up to your imagination: solid colors, other shapes, import parts of other pictures (maybe superhero symbols?), that kind of thing. Depending on the drawing software you use, you can do all the design now, or use the image-editing software needed for the next step.

When you've got your first torso design done, print it out on a piece of normal printer paper, cut it out, and place it on a minifig torso to make sure your sizing is correct. If not, make some adjustments and try again until you've got just the right size.

When you do have the size right, copy it as the basic template and then go about making more designs. You have a whole letter-size sheet of decal paper to print, and once you've printed something on it and cut it out, it's darn hard to use it again. So make a bunch of designs to work with.

We also will need a template for the face, which we'll make here, but we won't do the design work here. Faces should be about 1.5 in. long and 0.3 in. tall, so try making that as a rectangle, as a simple white shape (eventually, white will not be printed, so this is as good as a blank mask). Print on normal paper to test.

When you have your torso designs done, and your face template is ready, save them all as simple .png files. This type of file allows a color (usually white) to be set as "transparency," meaning it won't be printed.

Now we need another piece of software—an image editor. Adobe Photoshop Elements (or it's full-bore big brother) work very well on both Macs and Windows machines, though there are a number of good alternatives (Pixelmator on Mac, GIMP on Mac, Windows, and Linux).

If you have more design work to do on your torsos, open them in the image editor and proceed. Image editors make it easy to copy small pieces from one image to another, so it could be ideal for taking patterns or symbols from pictures you find on the Internet and adding them to your minifig decals. When done, save them, again, as .png files.

For the face decal, we're going to do something special. We are going to put your geeklet's (or your) face on the minifig!

We need to start with a picture, taken fairly close up and directly in front of your subject. The face should fill the frame as much as possible (so we don't have extraneous stuff to trim out at the end), and should be well lit so there are no shadows.

Since minifigs have black features on yellow plastic, we need to convert the photo to black-and-white using the tools in the editor, so that when the decals print, the white becomes transparent, and only the features will transfer to the minifig.

To start, in Photoshop, we use the menu sequence Enhance > Convert to Black and White. This gets us to a lovely grayscale image, but which still has a lot of midtone detail that is far too complex for a minifig face. So we want to pump up the contrast to drive out midtones. Use Enhance > Adjust Lighting > Brightness and Contrast. And then use Enhance > Adjust Lighting > Levels to play around with the tones until you are left with an image of the face in which the skin has gone to white and the features, like eyes, nose, and mouth, are mostly black lines. It should look very stylized and yet still be recognizable as the person in the original picture.

You can also use some blending to drop out all the fine details, leaving only the black outlines of features, or this could also be done by using an eraser or layer mask (for those with a little more practice on the software).

When you are happy with the features, you need to shrink the image to the size for the face decal. The image should be less than 0.2 in. high (0.3 in. overall height of face decal—the features should not be right at the edges of the decal). Again, print out some tests on regular paper, cut them out with your hobby knife (parents help younger kids with this), and make sure they will fit on the face of a minifig head.

When they're ready, save both your torso and face images as new .png files and set up a Word (or other word processor) document to the size of the decal paper (probably 8½ x 11 inches). Insert multiple rows of shirts and faces—either duplicates or make many variations. The torsos are especially easy to make in a number of colors and styles. There's no sense wasting decal paper, because you can't run a sheet through twice, so make the most of a print job. You should print at the highest resolution possible for the clearest, sharpest features. And again, test print first—decal paper is expensive!

After printing, you should let the page dry for at least five minutes so ink won't run when we soak it for the decal application. An optional upgrade is to spray-paint the printed decal sheet with clear-coat spray paint to protect the ink even more from the water during the transfer process. Be careful during this step to only lightly coat the paper—better to do three light coats, because a heavy coat will saturate the paper and ruin some of the decals.

STEP 2: Preparing the Minifig

Take apart your minifig, separating each section: legs, head, torso, and arms. Popping arms on and off is pretty easy; everything just snaps together. Separating all the parts will make working with the torso easier. And for future reference, this would also make it easier to apply decals to the arms or legs if you ever get that ambitious.

If your head or torso has ink printed on it, you need to use the Brasso (or equivalent) to remove the ink. This step should by done by the parent, or by the child with proper supervision and safety precautions. Brasso is pretty pungent stuff—make sure you have proper ventilation.

Work on an old rag or towel to protect your work surface. Lay down your printed piece and apply a small amount of Brasso to the corner of an old rag. Work the Brasso into the printed area. With a little elbow grease, it won't take long; perhaps a minute or less, and your piece will be free of printed ink. When all pieces are done, clean with soap and water to remove any extra Brasso, and make sure to clean your hands. Let everything dry.

STEP 3: Apply the Decals

Use scissors to closely cut out each of your designs along edges, and set them together (but not touching) on a paper plate or towel. One at a time, use tweezers to dip a decal in distilled water (using a small, low dish will make this easier). Dip only the decal, in and out of the water; do not hold under water for any more than half a second.

Set the decal on another, clean paper plate for one minute to allow the water to saturate the paper and decal and activate the adhesive. Use a small paintbrush to apply a very small amount of distilled water to the surface of the minifig piece where the decal will be applied (the head or torso). After one minute, gently slide the decal onto the minifig piece, using your fingers. The decal should now be slick and slip off the backing paper and onto the plastic surface easily. Discard the backing, and use your fingers to position the decal.

Use a cotton swab to roll out any air bubbles. Use a paintbrush to apply decal setting solution to help decal set and adhere. Set the figure aside to dry, and work through the rest.

Once the decals are completely dry, apply a coat or two of clear-coat spray paint to completely seal the pieces (using glossy clear-coat helps approximate a plastic look).

NOTE: None of the sealant steps are absolutely necessary, so you can omit them if you need to for the sake of time. However, they will help make the final product far more durable and long-lasting.

When everything is dry, reassemble your minifigs, taking care not to rub the decals directly. They'll be relatively sturdy but can never replace the original paint for durability.

Now you and your geeklet are ready to conquer the universe with your armies of crack minifig commandos. Or, you know, your own mini-mes!

Alien Drums

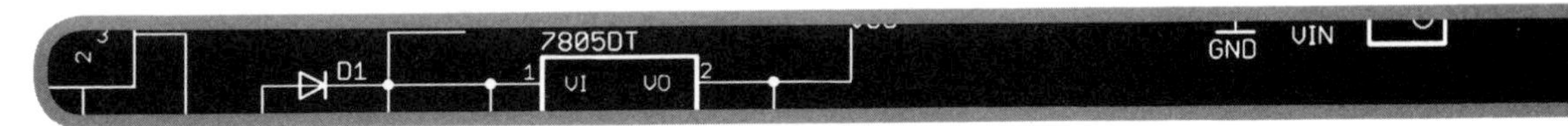

(Project Idea by Chuck Lawton)

A sage man once said, "I don't want to work, I want to bang on the drum all day." Truer words were never spoken.

And for littler kids who express themselves better through action than speech, well, the truth of that statement gets even "truthier," as Stephen Colbert might say. Music is a passion for many, especially many geeks (considering the match and science involved), and even when we don't have the technical skills to produce it, we all like to sing along in the shower.

But the good news is that we are in an age when even people who can't play instruments can make some really cool music just with software and a laptop. And many of those people making this kind of music are geeks (Who better to push laptop and software to its limits?).

But such tools are for older children and adults who understand computers, understand some manner of music theory, and have an idea how to marry the two for a specific outcome. For those who don't have even that kind of musical outlet, like our kids, what to do? Children don't start with all that knowledge. They start simply with passion and curiosity.

It's our job to give them the basic tools to feed that passion and curiosity. But what tools? Well, toy pianos and synths aren't much good without training. And neither are toy guitars without lessons. But drums. DRUMS! Well, anyone can get into playing drums!

But we are geeks and parents. Running out and buying our kids a set of bongos just isn't going to cut it. We don't get anything out of it, and to them, it'll just be another cheap toy they can play with once and then lose in the back of the closet. What if we could make acquiring the drums as important an experience as playing them? What if we could build the drums with them? Yeah, that would rock.

PROJECT	ALIEN DRUMS
CONCEPT	Use PVC plastic pipe to build a set of tuned drums like the Blue Man Group uses.
COST	$$–$$$
DIFFICULTY	⚙ — ⚙⚙
DURATION	☼☼ — ☼☼☼
REUSABILITY	🌐 — 🌐🌐🌐🌐
TOOLS & MATERIALS	Two 8-foot lengths of 2-inch-diameter PVC pipe, purchased at any home improvement store; 6 2½-inch PVC elbow joints; 2 2-inch PVC couplers for experimenting; one package of large zip ties; 3½-inch-length screws (optional); old pair of flip-flops

This project is relatively straightforward, and even a very young geek kid can help with it. The only safety issue to take into account is how you're cutting the PVC. If you can get the pieces cut by the helpful folks at the store where you purchased the materials, you're home free. If you're doing it at home, you need to take some care. It's quick and easy to use a chop saw and do it yourself (or if your child is slightly older, you can guide him or her through the process, with appropriate safety training). You could also make things a bit safer, and a bit more manual, using something like a hacksaw, which will still cut through the PVC pretty well. A younger kid can use it fairly safely, and learn a valuable skill or two.

What's even better is that you both can learn some fundamentals about the physics of sound by constructing these drums. The vertical tubes will resonate at a pitch based on the length of the tube. The drums can be tuned to be in various key signatures and, if built out chromatically (tuned to the chromatic scale so that each note plays well with the others), can be used like a xylophone to play songs and melodies. These are precisely what the Blue Man Group based their whole amazing career on. Indeed, maybe you can find a video to show your kid before building these, just to get them excited.

STEP 1: Cut the Pipe

Using either your chop saw or a hacksaw, measure and cut 16 lengths of the PVC pipe as follows:

- Four lengths of the PVC pipe at 25 inches. Using a Sharpie, mark one *Low C*, one *E*, one *G*, and one *High C*.
- One length of the PVC pipe at 18¾ inch. Using a Sharpie, on the inside of the pipe, write *Low C*.
- One length of the PVC pipe at 7½ inches. Label the inside of this pipe *E*.
- One length of the PVC pipe at 2½ inches. Label the inside of this pipe *A*.
- Three 3-inch lengths of PVC pipe. These will be used to join the elbows together.

When you're done, you'll have 16 pieces: 4 long sections of pipe, 6 elbows, 3 equal short cuts of pipe, and 3 variable lengths of pipe.

STEP 2: Assemble the Alien Drums

We'll start with the first tuned pipe. Take the 25-inch piece of PVC pipe (marked *Low C*) and put an elbow on one end. To the other end of the elbow, add one of the 3-inch lengths of pipe. Add another to the end of this short piece, and then the $18\frac{3}{4}$-inch length (also marked *Low C*) to the elbow, adjusted so it points back up parallel with the 25-inch pipe.

Using a flat surface large enough to cover the end of one pipe (but not much bigger)—perhaps something like a flip-flop—as a mallet, strike the top of the 25-inch Low C pipe so you cover the opening with the flip-flop as you strike it. You should hear a tone when you do this.

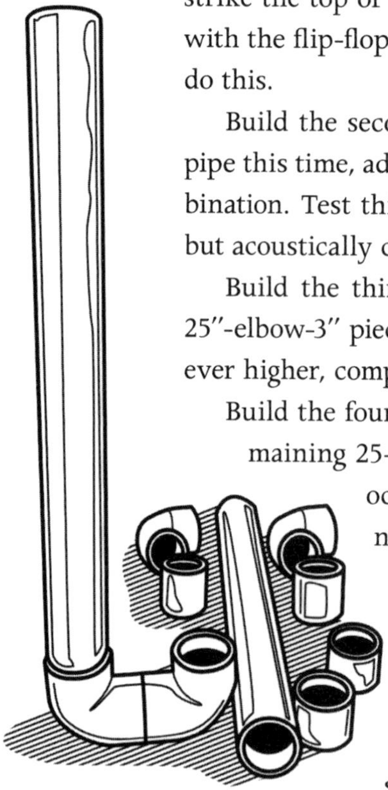

Build the second assembly using the 7-1/2-inch ("E") length of pipe this time, added to the 25-inch-elbow-3-inch piece-elbow combination. Test this one with the flip-flop, and you'll hear a higher but acoustically complementary tone. You are building a scale!

Build the third assembly using the *A* length of pipe and the 25"-elbow-3" piece-elbow combination. A third test will get you an ever higher, complementary tone in line with the previous two.

Build the fourth assembly by doing absolutely nothing. The remaining 25-inch PVC pipe is already tuned to high High C, an octave above the lowest drum (meaning it's the same note, just at a higher tone).

Striking the drums in order from the longest (the one with the $18\frac{3}{4}$-inch end marked *Low C*) to the shortest (the solitary 25-inch pipe) should produce a C-major chord. If some of the pitches are off, double-check your measurements and ensure that the elbows are pushed together tightly.

STEP 3: Finishing Touches

To assemble the four separate drums into one drum kit, gather them together so that each of the 25-inch lengths are upright adjacent to one another, creating a 2x2 grid to make hammering each one easy. Let the ends that produce the sound point away from one another in four directions, which conveniently gives the whole build support. Because the High C tube does not have an elbow, be sure to keep that tube flush with the top as you screw it in.

Bind together the lengths of PVC in the center core, using the zip ties to make the drums stable and portable. If the drums still have some looseness after binding them, you can use some 3½-inch screws drilled from one tube to another to prevent the drums from sliding up and down as they are struck.

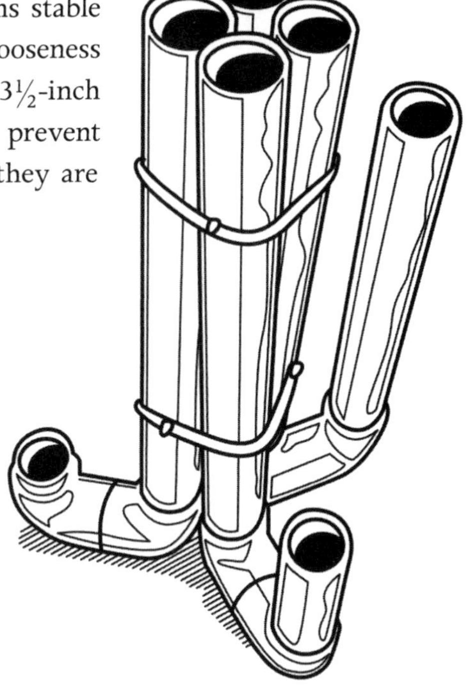

What's the Science Here?

The tubes produce a tone because when you strike one end of an assembly with the flip-flop mallet, you force air through. That air resonates at a particular frequency related to the length and diameter of the tube. In our particular assembly, Low C is 264 Hz, E is 330 Hz, G is 440 Hz, and the High C is 528 Hz. Pipe organs work in a similar way.

An Extra Geeky Idea

You can adjust the tone of a tube by adding length to the assembly. Use a straight coupler and PVC pipe cutoffs to make whatever pitch you like. Longer pipes will make lower pitches.

If you are more ambitious, you can build an entire 12-note scale. Or even 24 notes. Then use colored tape to mark each pipe/note in a chord in a given chord progression. With that, you can teach your geeklet how to improvise a whole song based on tapping the notes in one chord, and then in rhythm, moving to another chord, and another to form a tonal progression. At that point, it's not just banging on drums anymore, it's making music.

Create Stop-Motion Movies

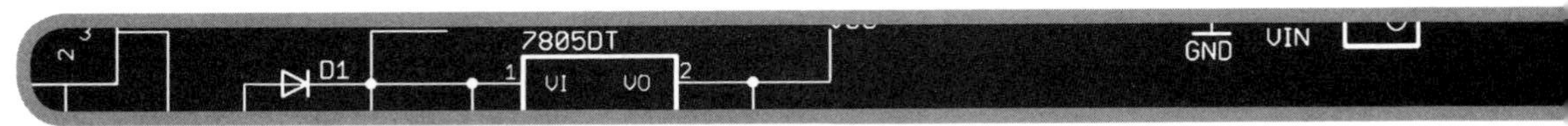

(Project Idea by Dave Banks)

The phones we carry in our pockets these days are capable of shooting high-definition video. The digital single-lens reflex (DSLR) cameras are good enough that even professionals are shooting episodes of television shows with them. Basically, there has been a perfect storm of technological development and good old-fashioned capitalist competition coming together in the last few years to advance what the average user can do when it comes to DIY video making.

So, many of us get the bug to create with video. But what can we do with it? And especially, what can we do with it with our kids?

Stop-motion animation has been a popular technique in film for decades. It was first invented to make objects appear as though they were moving magically—people gliding across the ground without walking, or imaginative moving line drawings. You see it used in the Claymation of *Wallace & Gromit*. As technology has gotten more sophisticated, the number of people trying stop-motion at home has exploded. Stop-motion lets you do things with video that you can't do with live action. For example, you can assemble LEGO kits, or move toys around and have them interact with one another, or make drawings come alive. Your kids can even become superheroes with amazing speed or even teleporting ability. This simple step-by-step project is a cool way to make a short stop-motion movie with your kids.

PROJECT	CREATE STOP-MOTION MOVIES
CONCEPT	Bring out your inner Spielberg by making stop-motion animated movies with a few simple materials, camera, computer, and your LEGO bricks.
COST	$–$$$
DIFFICULTY	⚙⚙ – ⚙⚙⚙
DURATION	☼☼☼ – ☼☼☼☼
REUSABILITY	🌐🌐🌐🌐
TOOLS & MATERIALS	Computer; camera; LEGO bricks **Optional Parts:** Tripod; poster board; masking tape; lights and lightbulbs or a store-bought light box

There is a project in my first book, *Geek Dad: Awesomely Geeky Projects and Activities for Dads and Kids to Share*, that explains how to create your own comic strips, using a digital camera and your LEGO minifigs or other action figures. The setup for that project is pretty close to identical with this one. Basically, you need your camera, something to take pictures of, and the right lighting setup to take those pictures effectively. Then you process those images, using your computer and one of a number of software programs available. The big difference here is that, instead of making a static series of images, you'll take many, many more pictures and string them all together to make something dynamic.

STEP 1: **Write the Script**

It's always important to have a plan, and for a movie like this, the script is your plan. So what kind of movie do you want to make? Will your story be original? Will you LEGOfy an existing scene from your favorite movie (*Star Wars* and *Raiders of the Lost Ark* have already been done, but how about some Muppets)? Try to come up with a visual gag or a joke, or come up with a clever approach to make your movie more memorable. You can even look to YouTube for inspiration—do a search on LEGO, and you'll bring up quite a few examples (maybe start with the stop-motion LEGO Batman & Superman Movie at http://www.youtube.com/watch?v=FMQsNQRaYoE).

The format for your script can be text, like the kind of scripts you see for TV and movies, but an even better idea is to make it a storyboard. A storyboard breaks out your movie, scene by scene, action by action. Use text to describe action, dialogue, and music for each scene. Sketch out how scenes should look. You don't have to be a great artist for this part: Stick figures are okay!

STEP 2: Build the Set

Once you have your story, you need to build your set and collect your actors. Sort through that huge bin of LEGO bricks you have to find the bricks and minifigures you need to act out your scene(s). You may have to rewrite or tailor your script to match the bricks you have. And I'll admit: It doesn't have to be LEGO. You can use other materials. Maybe your favorite action figure is the star. Or you have Matchbox cars zooming around. Or you could create some interesting creatures with Play-Doh. The possibilities are wide open. However, LEGO bricks make things very easy to put together and take apart, and there is something timelessly fun about them!

Build your sets and do a quick rehearsal of how you want your story to go. If you are using LEGO BRICKS, you can build your set on top of one of the large, flat LEGO bases that come with some kits. That way, the structures that are meant to stay fixed in place will do so. Otherwise, use a level, flat, neutral-colored surface that won't show scuff marks and is easily cleanable. Maybe a large cutting board. Make sure you have collected everything you need to tell the whole story before you get started.

And from a creative standpoint, think about embellishing your movie set with additional props. Attention to detail is what will push your movie from being "okay" to being great. Example: When the Dianoga monster periscopes his one eye up above the surface in the trash compactor scene in *Star Wars* (yeah, bet you didn't know it had a name!), this little detail is what makes the scene memorable. What can you add to your story that has nothing at all to do with the direct story you're telling but adds a bit of fun?

STEP 3: Set the Stage

When you make a movie, you need to keep in mind issues of continuity—making sure the things that should look the same from frame to frame do, for example. One element that can change over time is your lighting. Since your stop-motion movie may take a while to shoot, it's better to control the lighting yourself rather than rely on natural light, which can shift a lot over an afternoon.

It's better to set yourself up in a space without natural light. A garage would work, if you can cover any windows with, say, towels. Or a basement is also ideal. You can use cheap aluminum reflector lights with clamps on them, and "natural light" lightbulbs, to illuminate your set. If you set up two lamps shining from two directions, it will minimize most shadows. You can diffuse the light by hanging two sheets of tissue paper in front of the lamps (but don't let the paper touch the bulb, or you could get some burning). If you set up this way, you won't need to use the camera's flash, which would have a tendency to wash out your shots by overexposing them.

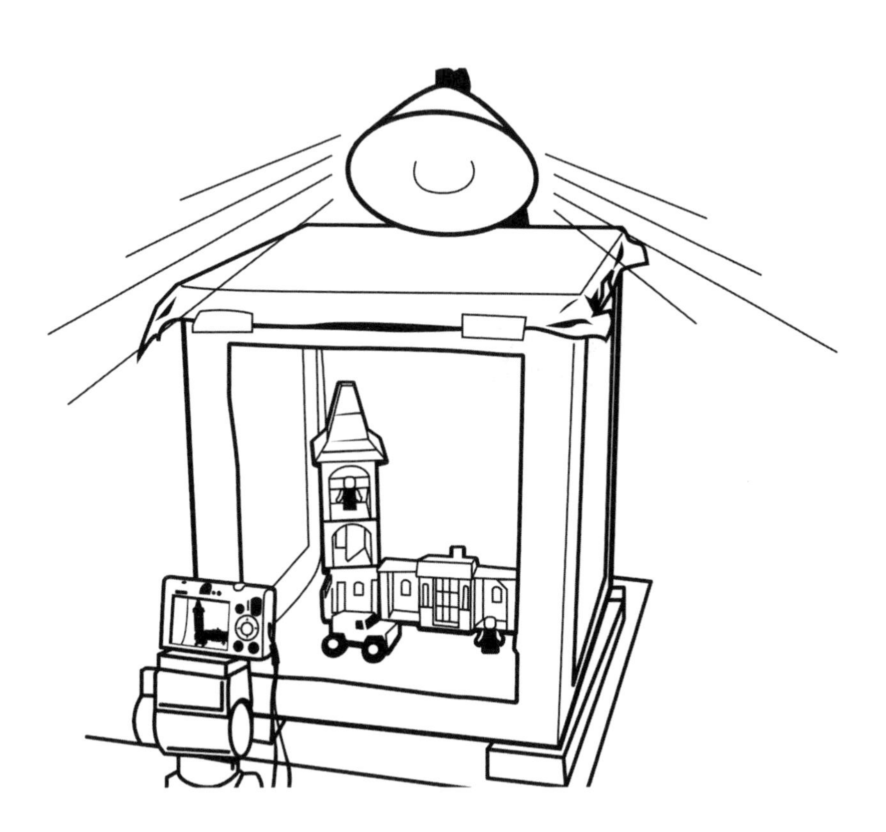

To neutralize any background, and to bounce light back on the LEGO BRICKS, use a piece of white poster board. Push a table up against a wall and lean the poster board halfway on the table and halfway up the wall. The poster board should have a gentle slope to it, like a skateboard ramp, so that if you looked at it through the constrained view of your camera, you couldn't see any creases or folds (imagine every movie version of a director walking around with his fingers shaped like a square, looking at everything). Use masking tape to fix the edges of the long ends to the wall and the table.

Place your set on the table on the poster board. Set your camera in its shooting position and check the frame. You want to see just your set and a neutral white background without edges or lines. You may want to take a photo or two to check lighting and to make sure everything looks how you want it.

OPTIONAL: There are also inexpensive light boxes available on the market to make this part of the process simpler to put together. They often come as foldout fabric boxes with clamp-on lights, which will create the same well-lit space with a neutral background.

It's vitally important to secure your camera. Moving the camera (or the set) between shots will cause discontinuity and thus cause the viewer to find the movie less believable. You can secure your camera by locking it down on a tripod or by attaching it to a table or counter with either masking tape or tack putty. If you use a tripod, even consider using masking or duct tape to secure it to the floor so you can't even accidentally knock it over between shots.

STEP 4: **Shoot Your Footage**

Now you can begin following your script. As you shoot your scene(s), consider having more than one element moving on your set at any given time (in any given shot)—it makes the scene more believable if more is going on than just the main character moving.

After taking your first photo, you should move each figure or brick only very slightly. Keep in mind that a movie you see at the theater uses 24 individual images (frames) per second to create the seamless live action you're used to. You don't have to go quite that far, but you should aim for 12–15 frames per second to make one second of video. Each frame is a separate photo, so to put that in perspective, you'll need as many as 900 photos for one minute of video.

This will likely take a little practice to know how much (or how little) to move your bricks, so you may want to consider doing a test shoot before getting started.

Be careful of shadows as you move about—check before clicking the camera's shutter. You want the set to look the same from shot to shot.

Move, click, move, click, move, click. Keep it up until you've completed your scene.

STEP 5: Making Your Movie Come to Life

Once your principal shooting is complete, it's time to move to post-production. This is where you make your photos move.

There are many options for the software you can use to get this job done. Some cost money, some are free. For your first stop-motion movie, consider using Apple's iMovie or the Windows Movie Maker. Slightly more powerful, Adobe's Premiere Elements is available for both the Mac and Windows worlds. Each brand of software is a little bit different, but the process is basically the same.

You need to import all the photos from your camera to your computer, using whatever process is appropriate for your camera. After getting them onto your machine, you need to import them into the moviemaking software. Windows has its own Movie Maker application, and Mac has iMovie.

Then bring the images, in order, onto the video timeline. Most cameras use a very basic file-naming convention when taking the pictures, so the numerical sequence of pictures should mostly follow the timeline in which you shot them, unless you had to go out of order. If you used the system of starting a sequence with a marker identifying the sequence, that will help you keep things organized now.

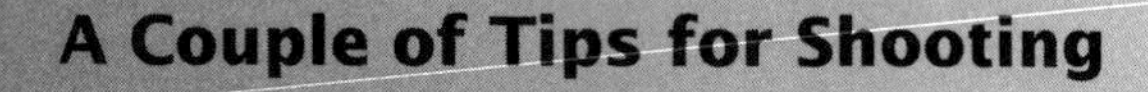

A Couple of Tips for Shooting

This is an energy-intensive process. Turning off your camera's flash will help, but you may want to have an extra fully charged camera battery on hand, or anticipate taking a break to recharge the camera.

Stop-motion is tedious work. Don't forget to take some breaks for you and your kids!

The memory space on your camera could become an issue when taking this many pictures. But one very important thing to consider is the resolution you're shooting at. Most digital cameras can shoot upward of 12 megapixels in the RAW format. That means each image is generally around 12MB. Trying to take hundreds of pictures at this size will really fill your camera up quickly.

Luckily, we don't need that kind of resolution for the kind of movie we're making. Realistically, we need images of less than 1-megapixel resolution. My little Canon PowerShot has a setting for 0.3 megapixels, which creates images 640x480 pixels. This is good enough (and will be a lot easier to work with when it comes time to assemble all the images into a video).

You can get creative here, too. If an action is repeated in your script, don't photograph the image more than once. Simply copy and paste the animated images during postproduction and reuse the footage you've already taken once.

These instructions also assume you're shooting your script in sequence. However, if you want to work out of sequence for some reason, or you can guess that you may need to go back later and reshoot some scenes, you may want to initiate a system not unlike the clapper boards used in "real" moviemaking. Use a piece of paper and write down the name for a given scene or motion sequence. Take a picture of the paper before you shoot the scene so you have this marker in front of the series of image files on your camera when you come back to work with them in postproduction.

Select all the images in the timeline and set the duration so that each image will show to approximately ½ second. Play the movie as a rough cut to make sure it looks how you wanted and that all the images are in the proper order. At this point, you can play with timing to get the motion smooth and get the proper visual beats you want.

Once you are satisfied, you can add music for a soundtrack (RIAA-legal only if you plan to upload this to YouTube), sound effects, or to record dialogue for your characters. You may need to go back and play with the timing to line some things up with your audio. And then export to a movie file from your software. You'll want to use some kind of fairly common file format—.avi or .mp4—depending on your moviemaking software.

Then all you have to do is copy the video to a file, depending on the software you're using and where you want to present it (on your computer, on your phone, on your TV, again depending on the technology available to you). When you're done, gather the family, pop some popcorn, and enjoy your masterpiece! Cannes can't be far behind.

Easy Electronic Music

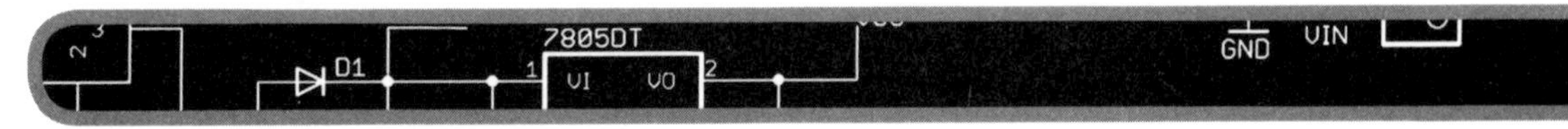

Special Educational Segment by Z from www.hipsterplease.com

Way back in January 2008, I was just through my first year as editor of GeekDad.com, I saw a picture one day of an awesome tattoo posted on www.boingboing.net—a great geeky site, by the way. Some guy in the South had gotten a d20 (the most important die in Dungeons and Dragons) tattooed on his bicep. It was sublime, and yet evocative. It was the ultimate geek tattoo. I posted about it on GeekDad.

Through some combination of events, I invited the fellow, known as Z on his nerdy music Web site, hipsterplease.com, to join one of our podcasts. He was a great guest and also, it turned out, a dad. He had some measure of notoriety in the chiptune and nerdcore hip-hop music scenes (yes, there are such things) for his own podcast and championing of the genres, so I wasn't sure if his plate was already full. But after our podcast together, I invited him to come write about music, and whatever else he wanted to, on GeekDad. To my joy, he agreed.

Three years and nearly one hundred podcasts later, Z is still my co-host for our a biweekly HipTrax *podcast, where we feature geeky music you can share with your kids. He is also one of my Core Contributors, and the person I know who has the deepest knowledge of music that is outside the mainstream.*

With the advances in music technology over the last few years, making your own music—especially making it with your kids—is easier than ever. So I asked Z to write up a primer for us geeky parents on the tools available (many for free or very cheap) to make some very cool and retro video-game-inspired tunes.

PROJECT	EASY ELECTRONIC MUSIC
CONCEPT	Use specialized software and tools to make music like we used to have in the video games of our youth.
COST	$-$$$
DIFFICULTY	⚙ — ⚙⚙⚙
DURATION	☼ — ☼☼☼☼
REUSABILITY	🌐🌐🌐🌐
TOOLS & MATERIALS	See below.

Chiptunes refers to music that is created using the sound chips of often antiquated computers or video game systems, Remember how the music sounded on your Atari 2600 or Nintendo Entertainment System (NES)? Those were the early days of sound reproduction on video game machines, and the tools for creating music were primitive. Tones had a very electronic feel to them, and music could often be pumped out only one note at a time (no chords). Nowadays, there is a movement to create music using the sound modules—the chips— from those old devices as the primary "instruments." Hence *chiptunes*.

Typically, the term itself now conjures images of powerful dance music blasting from a cunningly manipulated Nintendo Game Boy (www.nullsleep.com), or the sound of an NES (Nintendo Entertainment System) chip being pushed to its limits atop a more traditionally rooted rock performance (www.ifightdragons.com). Still, while it may seem as if Nintendo products have cornered the chiptune market, the truth is that practically every old gaming system and electronic toy imaginable has been compromised in the name of music by some dedicated modder (a hacker who specializes in retro gaming systems).

In a sense, chip music is the perfect metaphor for modern nerd life, since it combines nostalgia, technology, creativity, and, best of all, experimentation. So it is the ideal activity outlet for a Geek Dad and his brood.

As counterintuitive as it may sound, because the chips are so limited in their ability to produce sounds, the strength of chiptunes is the style's unique flexibility—there is no right or wrong way to approach its creation. There are a myriad of everyday tools and programs available for the chip musical artist. A motivated Geek Dad need only look around his own house to find a suitable instrument for chiptune-making mayhem, perhaps an old Game Boy, or a Casio keyboard. But, if you've already sent all your classic game and computer systems to that great scrap heap in the sky, you can substitute more modern components to achieve that all-important retro sound.

Here's a trio of ideas to help you get the ball rolling.

JOHAN KOTLINSKI'S LITTLE SOUND DJ:

Affectionately known as LSDJ (www.littlesounddj.com/lsd), Little Sound DJ has been a defining force over the past decade of Game Boy–based chip music. This tiny sequencer pushes four channels of 4-bit sound, and, though it's no longer available as a proper cartridge release, a ROM (Read-Only Memory) of the current build will set you back a mere $2 U.S. (ROMS are pieces of software that can play on older game systems—they only need to be copied onto blanks of the original chips or cartridges used by those systems.)

This ROM image can be loaded to a Game Boy, Game Boy Advance, or DS flash cartridge for use as it was originally intended, but those not interested in investing in specialized (and, in some countries, illicit) tech can simply take advantage of the power of desktop emulation. There are a number of free Game Boy emulators, such as VisualBoyAdvance (http://vba.ngemu.com), freely available for download. You can load Little Sound DJ into such an emulator, just like any other game ROM, for instant chiptune action. The only barrier that exists is the learning curve inherent in the software itself.

The visual interface is minimal at best, and understanding the intricacies of a sequencer like LSDJ can be daunting. If you've never sequenced a song before, you will likely be intimidated by the series of digits, dashes, and abbreviations that greet you on start-up. But fear not. Not unlike coding for a Web page or other software, every character represents a part of the music. With some trial and error, you and your kids can pick up the new "language." The bottom line of what you're doing is that you are arranging notes across a timeline. LSDJ provides an amazing amount of depth in how to make chip music in the Game Boy style, and there's a solid online community full of dedicated artists who regularly share their knowledge.

Once you've managed to bang out your first creation, all that's left is recording the audio. For Mac users, this involves capturing playback through an application like GarageBand. If you have a PC or Linux, you can accomplish the same thing using a freeware recording application such as Audacity. You'll be able to plug external chiptune devices into your computer by running a cable from the headphone port on the device to the mic input of your computer, and recording directly.

COMPUTER + CASIO KEYBOARD:

While "Casiocore" (music from old-fashioned electronic keyboards) isn't exactly proper chiptunes, you can easily ape the lo-fi aesthetic of classic console, using any number of older electronic keyboards. As a rule, the cheaper the keyboard, the more lo-fi and "chippy" it sounds. So scrounge around your local thrift stores or pillage your children's toy instruments to build your minimalist musical armory.

If you are a high-level tech-minded Geek Dad, you may even want to try your hand at circuit bending, or creatively short-circuiting simple, solid-state devices (computer-chip-based hardware without moving parts) to create unique noisemakers. By doing this, you could turn anything from a toy piano to a LeapPad Learning System to a Speak & Spell into a one-of-a-kind electronic instrument by adding electronics, or hacking the ones already built in.

The advantage of this kind of music making is clear—by relying on devices with real, tangible keys and buttons, you open up a whole new world of performance opportunities. Turning what would otherwise be structured music-making into rules-free play is also a fantastic way to involve your children.

You can record the music on your home computer, using only a regular male-3.5mm-to-male-3.5mm cable (for devices with headphone or amplifier outputs) or by connecting the integrated speaker to a microphone plugged into your sound card's input.

IPHONE + NANOLOOP:

My personal favorite modern configuration for making chip music is Oliver Wittchow's Nanoloop (www.nanoloop.com). While the original Game Boy and Game Boy Advance versions of this application are still available in cartridge form, they cost a pretty penny ($75 and above). This handy little synth/sequencer, however, is now available for your iPhone for a mere $2.99!

While the visual interface is also fairly minimal, its innovative, intuitive touch controls make it a joy to noodle around with. Nanoloop supports six channels of synthesizer or sampler audio, an easily understandable loop function, and WAV audio file import (via e-mail) and export (to Mac OS X). Nanoloop's primary step sequencer is a 4x4 grid, with a separate grid for each of six instrument voices. Adding notes is as simple as touching the appropriate box (or step), and you can change notes by sliding your finger up or down.

You can access four individual grid banks at the top, and easily choose whether to loop the notes into a single grid or across all four. The touch controls also allow easy manipulation for everything from note volume to tempo, making Nanoloop a must-have for everyone from dedicated electronic musicians to those of us who just like to play around on our daily commute.

While WAV export is available only in beta, for those running OS X (at the time of this writing), capturing your audio isn't exactly a chore for the rest of us. Just plug the headphone output of your iPhone into the line-in connector of your computer (via that run-of-the-mill male 3.5mm to male 3.5mm cable) and record playback via GarageBand, Audacity, or your sound recorder of choice.

Geeky Art Prints

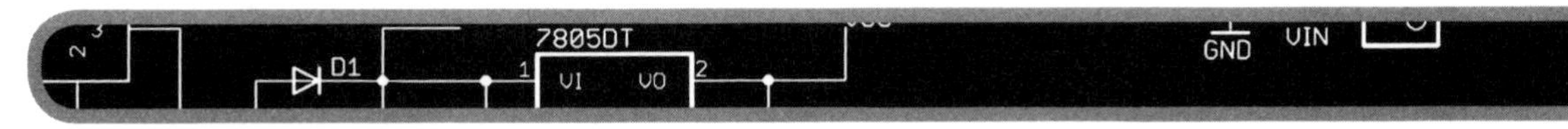

(Project Idea by Matt & Jenn Blum)

A geek's heart appreciates art as much as the next person. It just so happens that the works we favor tend to be more similar to the artwork of Frazetta (Frank) and Lee (both Jim and Alan) than Raphael and Donatello (the art guys, not the Ninja Turtles). But art also has a significant place in the imaginative world of the geek. We appreciate the symbology of superhero sigils and the insignia of the armada in alien battles. So it makes sense that we should encourage our little geeklets to start developing a love of art early.

PROJECT	GEEKY ART PRINTS
CONCEPT	Create highly colorful and geeky art with shaving cream and food coloring.
COST	$-$$
DIFFICULTY	⚙
DURATION	☼ — ☼☼
REUSABILITY	🌐
TOOLS & MATERIALS	One can of foam shaving cream (gel will not work); one box of assorted food coloring; Popsicle sticks, toothpicks, or kabob sticks; 5x7-inch matte card stock or poster board; one squeegee scraper; a cookie sheet or other cleanable work surface

This is a simple art project that creates brightly colored prints out of some interesting household materials. It's perfect for primary school–aged kids, because the steps are very straightforward, and the results are striking and fun. But you might be inclined to argue whether it's really a Geek Dad–worthy project.

This project is geeky for three key reasons. First, the materials are geeky. Actually, it's the *hacking* of the materials that is the really geeky part. The very idea of using shaving cream—the apex of the DAD substance—as an art medium is a really fun thing. Plus using food coloring as your paint is always fun.

Second, it's hacking something historical. The original version of this art form, which is a mix of printing and abstract painting, was especially popular during the Victorian era. You can find examples of marbled paper on the inside cover of old books (and see that tradition carried through even into the twentieth century). But the old-fashioned method involved floating an oil-based paint on top of a pan of water.

Third, what you can make with these materials is geeky. You can create very striking abstracts that look like tie-dye on paper, or bright, beautiful marble. And with a little practice and imagination, you can refine your skills to create figures, shapes, and symbols. Or maybe even badges for your own superheroes or alien armies!

STEP 1: You want to work on a surface that is smooth and can easily be cleaned if any food coloring spills on it (you don't have to worry much about the shaving cream—it's basically soap; whatever it spills on will be cleaner when you're done). Squirt shaving cream onto the surface, making a rectangle roughly the same size as the 5x7-inch card stock you're using. The entire area should be covered with foam, like a layer of snow.

STEP 2: Add drops of food coloring onto the shaving cream. Use just a few, or many colors scattered over the entire area, however you like.

STEP 3: Use a toothpick or Popsicle or kabob stick to swirl the colors around, but don't mix colors together (you end up getting mud that way). Gentle swirling makes a design.

STEP 4: Place a piece of card stock on top of your shaving cream design, laying it as evenly and flatly as possible. Pat on it to make sure it is firmly soaking up the shaving cream. Count to ten slowly.

STEP 5: Pull the card back up again, lifting from two sides, evenly. Obviously, it will be covered in shaving cream and dye. Turn it over and lay it on the table, cream side up.

STEP 6: Use your squeegee to carefully scrape shaving cream off with one stroke across the card, revealing the results of your design!

And there you have it. Your first attempts will likely be abstract (though beautiful, of course!), and smaller kids will probably be happy with that. However, in time, with some patience, you should be able to achieve some far more sophisticated designs!

Photoshop Your Kids into Their LEGO Kits

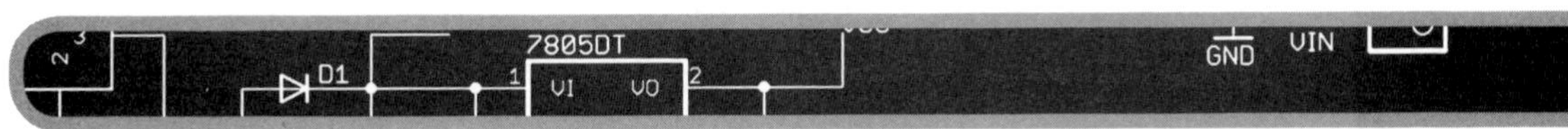

Celebrity Geek Dad Project by Ken Jennings

Answer: The greatest Jeopardy! *champion of all time, and also a very geeky dad.*

Question: Who is Ken Jennings?

Geeks certainly do have a proclivity for remembering trivial information, so it makes perfect sense that Ken Jennings, the man who won seventy-four straight matches on the television game show Jeopardy! *to become the all-time champion, considers himself one (well, he's also a software engineer). But even better, he's a geeky dad and enjoys doing interesting projects with his kids, including the project he shares with us below, using Photoshop image editing software to take pictures of your kids and make it look like they're playing inside their own toys.*

PROJECT	PHOTOSHOP YOUR KIDS INTO THEIR LEGO KITS
CONCEPT	Digitally shrink your kids and put them inside images of their own toys to make great greeting cards and portraits.
COST	$ (assumes you have everything already)
DIFFICULTY	⚙ — ⚙⚙⚙ (It's easy for kids to sit for pictures, and you can teach older kids how to do it all for themselves.)
DURATION	☼☼ — ☼☼☼
REUSABILITY	🌐🌐🌐🌐
TOOLS & MATERIALS	Digital camera; computer; Photoshop (or similar program); LEGO kits

When I was a LEGO-obsessed seven-year-old, my fondest dream was to live in a world made entirely of life-size LEGO bricks. Yes, the floors would be a little uneven and the chairs a little uncomfortable, covered as they would be with rows of round bumps. But it would be worth it, I was convinced, as I looked at the large-scale landmarks and other LEGO models exhibited in toy stores and LEGO theme parks. The real Mount Rushmore in South Dakota was fine, I guess, but the LEGO version of Mount Rushmore, built in Denmark in 1974 out of 1.4 million beige bricks—now that was an engineering marvel.

Well, my son Dylan is now a LEGO-obsessed seven-year-old himself, and I'm overjoyed that we live in an age when technology has finally made it possible for our kids to play in life-size LEGO models of their own creation. I'm not talking about Rick Moranis's shrinking ray in *Honey, I Shrunk the Kids* movies, which is, tragically, still unavailable for home use. Instead, in our house, we just use Adobe Photoshop.

To give credit where it's due, this is an idea I stole from the brother-in-law of my brother-in-law. (This is not one of those brainteasers in which I'm referring to myself. Your brother-in-law's brother-in-law can also be an entirely separate person, as a moment's thought will confirm.) My brother-in-law's brother-in-law, Duncan (nerdy name not fictionalized for this story!), likes to Photoshop pictures of his kids into tiny settings. His first three kids were little girls, so most of his miniaturized photography is in the fairy princess milieu: giant pinecones and dandelions and autumn leaves and so forth. But I took one look at his cobwebby oeuvre and had just one thought: This is how you put your LEGO-loving kids inside their LEGO sets!

The process is simple. Take a photo of your kid's latest LEGO creation, pretending to listen and nodding vigorously as s/he describes to you what all the parts do, how the ViewScreens and torpedos work, etc. Then take a second photo of your kid or kids in an appropriate pose. Cut out and overlay the child's picture into the LEGO photo, using the photo-editing software of your choice, and voilà!

Finally, a portrait that the grandparents will politely decline a copy of but your kid will stare at for hours.

Here are a few tricks I've discovered that help make the finished product as convincing as possible.

- Be sure to take the photo of the LEGO creation and of your kid at the same time and from the same angle, so the lighting will match. If possible, position yourself in the same spot, and take both pictures along the same line of sight.
- In the LEGO close-up, make sure your focus is as sharp as possible, since even slight blurring will keep the kid photo from blending in well. Use shallow depth of field (or a macro mode, if your camera has it) to give the impression that your kid has been shrunk to Lilliputian proportions; deeper focus will give the impression that he's life-size, but his LEGO bricks are huge!
- Take many kid photos, at slight variations of angle and pose, to increase your odds that one will smoothly fit the LEGO-scale photo.
- Once your child's photo is overlaid into the LEGO photo, soften the hard edges around its silhouette. Most photo-editing suites have "Soften" or "Gaussian blur" tools that work well for this.
- If you are confident in your "Photoshop Fu," consider adding shadows to really help sell the illusion.

My kids love seeing themselves at LEGO minifigure scale. In fact, when the final photo comes off the printer, the resemblance is complete because I see a wide minifigure-like smile on every face. You might even be tempted to make every family photo a LEGO Photoshop job! Just think of all the money you'll save on vacation trips to Mount Rushmore.

Death Star Temari Ball

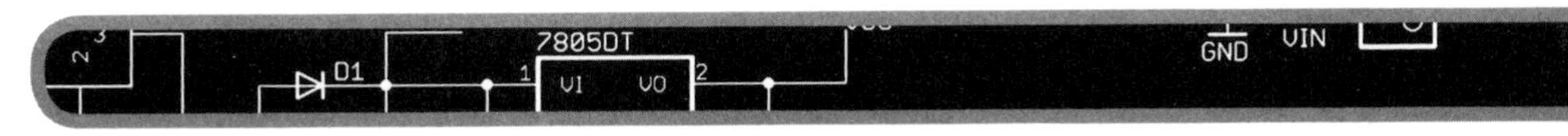

(Project Idea by Jonathan Liu)

So, what's a temari ball, you ask? A quick visit to Wikipedia will give you all the deets, but the short version is that they are both art and toy. The Japanese started by making them out of pieces of old kimono fabric, wadded up and wrapped tightly. Embroidery would be added to the outside in exotic patterns. If the ball was being given by a parent to their child (often in celebration of the New Year), the parent would put a secret wish for the success of their child in the middle of the ball. The balls are also often given as tokens of friendship and loyalty. Like so many traditional Japanese crafts, becoming a master takes training and dedication.

Sounds pretty neat, eh? But what has that got to do with geekery? Well, I'm here to tell you! I'm going to show you how to pervert a perfectly lovely and ancient oriental craft into something geeky. Think two simple words: *Star Wars*.

What important part of the *Star Wars* Universe is shaped like a sphere? That's right, the Death Star. You can use the traditional art of temari to make the coolest Death Star ever!

PROJECT	DEATH STAR TEMARI BALL
CONCEPT	Make a fully functioning temari ball, capable of wiping out whole planets.
COST	$$–$$$
DIFFICULTY	⚙⚙⚙ — ⚙⚙⚙⚙
DURATION	⏱⏱⏱ — ⏱⏱⏱⏱
REUSABILITY	🌐🌐🌐🌐
TOOLS & MATERIALS	**Core:** *The easy shortcut:* A wrapped mari core, in gray or black, available from www.etsy.com/shop/temarikai (the most comprehensive source in the United States) *The more crafty path:* Foam core ball 2½ to 3 inches in diameter OR a balled-up old sock or scrap of fabric roughly 3 inches in diameter (the foam core ball makes it easier to achieve a perfect sphere, but the sock/fabric is cheaper and more challenging); yarn, the softest you can get, any color; gray thread for base. Try a serger thread, which comes in a big spindle *(You don't need the yarn or thread if you use the prewrapped core above)* **Embroidery:** DMC Pearl Cotton size 5 embroidery floss in gray (slightly different from the thread base) and black (more if you want more variation) **Other supplies:** Embroidery needle, about size 14; straight pins; thimble; sewing scissors; small tape

Obviously, producing a single temari ball is somewhat a solo activity, at least in executing the primary tasks. You have two ways to go, depending on your kids' ages, if you want to make this a parent/child activity: Either both of you do your own temari ball, or you can act as the assistant while your child does the crafting. Both approaches can be just as fulfilling, as long as everyone works together, and occasionally hums the Emperor's March in harmony.

STEP 1: **Create the Mari (or Core) of the Ball**

If you are not using a premade core, this is your first step of the project. If you're using the balled-up sock or ball-of-fabric technique, make sure your lump is bunched up tightly until it's roughly a round shape. This is going to save you pain later. You can avoid this concern by using the foam ball.

Wrap the foam ball or sock/fabric with yarn, starting by holding down an end of the yarn on the core with your thumb and circling around a couple times to overlap it. Then start to go nuts, wrapping the core around and around, changing the direction of the wrap and squashing it frequently so it is as round as possible. If you've ever seen a baseball or golf ball without its outer shell, you know what we're looking for. Keep wrapping until the core is completely covered with yarn.

With the rough ball created with the yarn, which should cover the entire inner core so that nothing but yarn is visible, we'll now repeat the process with the gray thread. Where the yarn stops, overlap it with the start of the thread and, holding both in place with your thumb again, start to wrap the core, or mari, with the gray thread: Again, change directions often and continue to wrap until none of the yarn is visible. This takes a while, so you can do this while watching *Star Wars* for inspiration.

After the mari is completely covered with the thread, we need to secure the end of the thread so it doesn't stick out and cannot unravel. Cut the thread from its spindle with 4–6 inches of slack and use a needle to stitch zigzags across the surface until all loose threads are held tight. You now have the mari!

STEP 2: **Mark the Mari (Find Your Bearings)**

Using straight pins, mark a north and south pole on exactly opposite sides of the ball. You can eyeball it to start, but use the tape measure to check that they're spaced properly. Symmetry is vital for this project! Stick the pins in about half their length, and if you're right on, it'll look like there's one pin sticking all the way through the ball.

Now that you have your poles, you need to find and mark the equator! Using your tape measure, place six pins around the equator of the mari: Mark them accurately halfway between the north and south poles, and then measure around the equator to space them apart properly. Each pin will have a partner directly through the core of the mari. It'll almost look like we're making a model of Pinhead. . . .

Now you'll use the embroidery floss to mark the longitude lines and the equator. With floss threaded on the needle, push it into the ball close to the north pole, under some threads and through so that it comes out at the north pole, and pull your thread out far enough that no thread is visible where your needle entered. Then wrap the floss around to the south pole, crossing the equator just to one side of an equator pin (east or west doesn't matter, though you'll want to be consistent throughout the project).

STEP 3: Track Down Floss

At the south pole, catch a few threads of the core with your needle to tack your floss down, and continue around the mari, again crossing next to one side of the pin on the opposite side of the mari, and back to the north pole. Again, make sure that each time you pass on a side of an equator pin, it's on the same side as it was for the first pin. That is, if you started by passing on the west side of a pin, you always pass on the west side of a pin.

At the north pole again, go under a few threads to tack the floss down, and come out headed for the next equatorial pin. Repeat the first pass next to this next pin, through the south, and back to the north again. Repeat once more and finish at the north pole.

To finish off the floss, tie a knot into the end close to where it should finish. Insert the needle at the finish spot, and then back out somewhere nearby. Pull tight so the knot is sucked under the surface of the mari, where it will lock into place among all the intertwined fibers. Snip your exit floss close to the mari, and push the end back under the surface with a pin.

When you're done, it will look like an orange that has been

marked for cutting into six wedges (not to be confused with Wedge Antilles). The longitude lines are your guidelines.

STEP 4: **Embroider the Design**

Starting near one of the poles, you're going to stitch the floss around the ball from guideline to guideline in concentric "circles," sort of like stringing lights on a holiday tree. Except that they won't be circles, because between any two guidelines, the floss will sit in a straight line. This will create a visual effect that looks like hexagons when viewed from above, or spiderwebs. From the side, your lateral stitches will be like ladder rungs running up and down the six wedges of the ball (see illustration).

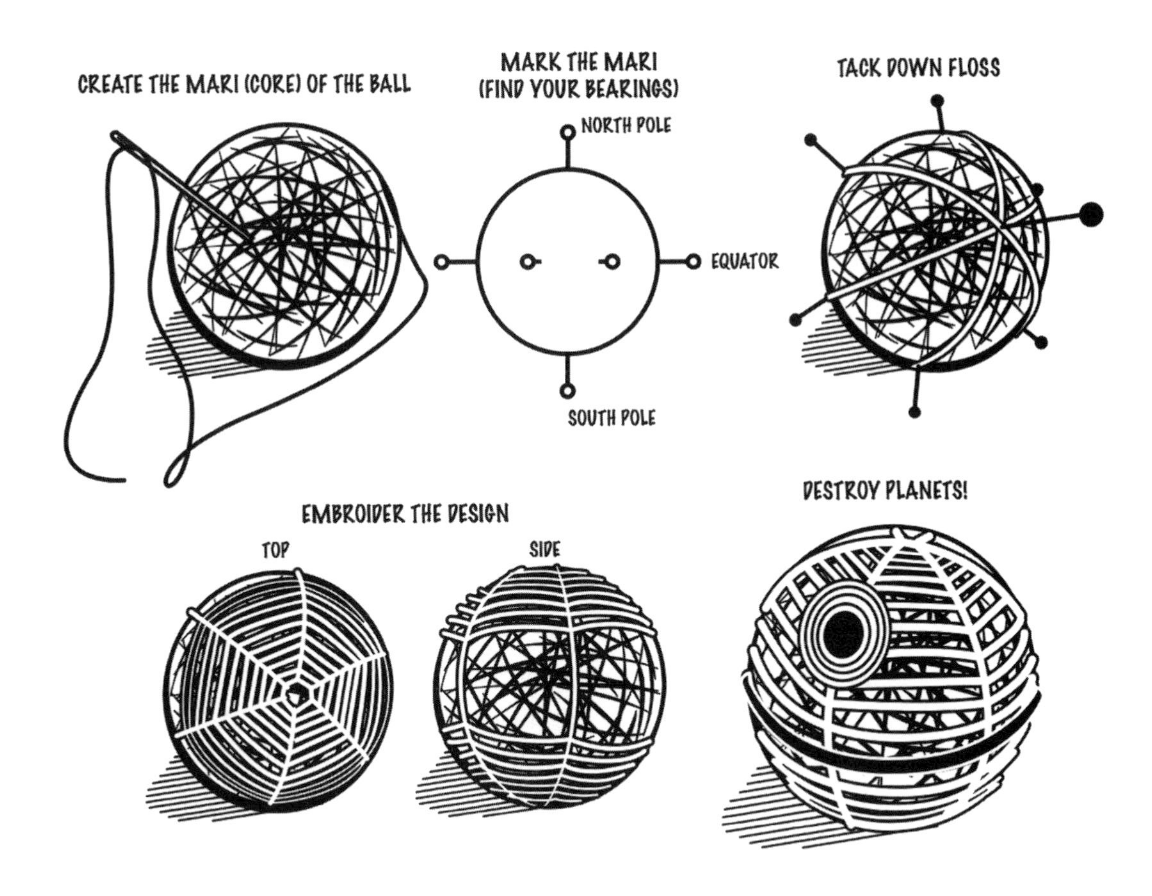
CREATE THE MARI (CORE) OF THE BALL
MARK THE MARI
(FIND YOUR BEARINGS)
NORTH POLE
EQUATOR
SOUTH POLE
TACK DOWN FLOSS
EMBROIDER THE DESIGN
TOP
SIDE
DESTROY PLANETS!

Start about a quarter inch from a pole, with your thread at one guideline, and then stitch laterally to the next guideline. Insert your needle under the guideline to catch a couple threads of the mari, come up the other side of the guideline, then loop back over the guideline and back under again to tie it off. Continue laterally to the next guideline, and so forth, until you've gone around the mari at that quarter-inch latitude and finished where you started. Start the next one another quarter-inch down toward the equator, and repeat the sequence around the ball at this latitude. Continue with these lateral lines until you've come within a half inch of the equator. Finish off the floss with the knot-lock method described above.

You can do all this by eyeball if you like, but if you want to be especially precise, measure along the guidelines and mark with pins so that your hexagons are evenly spaced. Once you've done one hemisphere, you should repeat, starting at the opposite pole.

Next, we are going to use the black floss to represent the trench. Insert the needle so the thread starts somewhere on the equator, then wrap around, keeping the floss neat as you go. (It helps to stitch into the surface of the ball from time to time, to hold the floss in place.) Finish off with the knot-lock and hide the end under another line of floss by simply stuffing the tail underneath some other threads.

Pick a spot about a third of the distance between the equator and a pole and, using the black floss, start a small spiral around that spot to represent the superlaser (you know, the big gun that blew up Alderaan). Start by inserting at the center point and bringing the thread out, and then catch a thread a very short way out nearby. Go under it, loop back over it, and then back under and up to loop it so that you've essentially wrapped the black floss around it completely. Then turn the mari slightly and work another loop around another nearby thread. When completing a loop, slowly turn the mari as you go, and do another loop, and another, building a trail around the center point so it becomes a black circle, about a half inch in diameter (but match the scale with the size of your mari). Finish with the knot-lock.

You can then switch to gray floss to build out the look of the superlaser dish another half inch or so. You can also add spokes, depending on how detailed you want to be, and if you wish, you can stitch other squares and rectangles onto the surface for even more detail.

Extra Geeky Idea

If you prefer *Return of the Jedi*, use black thread to create patterns on the surface to represent the incomplete portions of the rebuilt Death Star. I'll leave this for you to figure out. But for now, you can start keeping the larger systems in line with a little show of force from your finished battle station.

If you want any more info about temari, check out *Temari*, Barbara Suess's Japanese Temari book, as well as the Web site http://temarikai.com.

APPENDIX

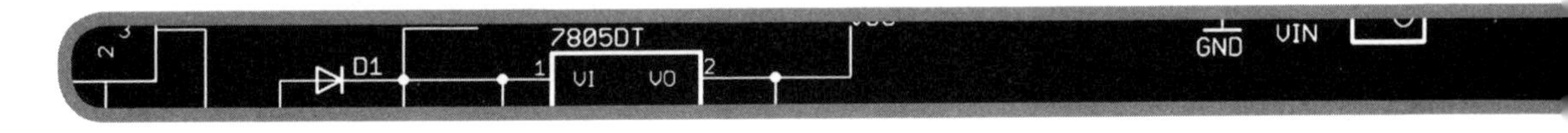

PROJECT	COST	DIFFICULTY	DURATION
HACK IT, BUILD IT, PLAY IT			
NERF Dart Blowgun	$–$$	⚙–⚙⚙	⏱–⏱⏱⏱⏱
Twenty-First-Century Superhero Cape	$–$$$	⚙⚙–⚙⚙⚙	⏱⏱–⏱⏱⏱
Trebuchet with LEGO Bricks	$–$$	⚙–⚙⚙	⏱⏱⏱
Hack Your Own Sound Box	$$	⚙⚙⚙–⚙⚙⚙⚙	⏱⏱⏱–⏱⏱⏱⏱
Go Medieval at Home with Make-Your-Own Weapons (by Chris Anderson)	$–$$	⚙–⚙⚙	⏱⏱–⏱⏱⏱
Smartphone Steadicam	$–$$	⚙–⚙⚙	⏱⏱–⏱⏱⏱
Skitterbot!	$$–$$$$	⚙⚙–⚙⚙⚙	⏱⏱⏱–⏱⏱⏱⏱
Make Your Own Ray Gun	$$	⚙⚙⚙	⏱⏱–⏱⏱⏱
Programmable Light Strings	$$–$$$$	⚙⚙⚙⚙	⏱⏱⏱–⏱⏱⏱⏱

PROJECT	COST	DIFFICULTY	DURATION
GAMING THE SYSTEM			
Make Up Your Own Combat Card Game	$–$$	⚙–⚙⚙⚙	⏱–⏱⏱
Fantasy Gaming Terrain	$–$$	⚙⚙–⚙⚙⚙	⏱⏱⏱–⏱⏱⏱⏱
Pun Wars	$	⚙–⚙⚙⚙⚙	⏱–⏱⏱⏱⏱
High-Tech Treasure Hunt	$	⚙⚙⚙–⚙⚙⚙⚙	⏱⏱⏱–⏱⏱⏱⏱
Pokémon Bingo	$	⚙–⚙⚙	⏱–⏱⏱
Backyard Zip Line (by Jamie Grove)	$$$$ (about $300)	⚙⚙⚙⚙	⏱⏱⏱
EAT, DRINK, PLAY, GEEK			
Igor Bars (by John Kovalic)	$–$$	⚙–⚙⚙	⏱⏱–⏱⏱⏱
Measure the Speed of Light with Chocolate	$–$$	⚙–⚙⚙	⏱–⏱⏱
Toy Candy Molds	$$–$$$	⚙–⚙⚙	⏱⏱⏱–⏱⏱⏱⏱
Dry Ice Ice Cream	$–$$	⚙⚙–⚙⚙⚙	⏱⏱–⏱⏱⏱
Homemade Root Beer	$–$$	⚙–⚙⚙	⏱⏱⏱⏱
CRAFTY LIKE A GEEK			
Clothes Hacking (by Patrick Norton and Sarah Holm Norton)	$–$$	⚙	⏱⏱
Make Your Own Mini-Me	$$–$$$	⚙⚙–⚙⚙⚙	⏱⏱⏱–⏱⏱⏱⏱
Alien Drums	$$–$$$	⚙–⚙⚙	⏱⏱–⏱⏱⏱
Create Stop-Motion Movies	$–$$$	⚙⚙–⚙⚙⚙	⏱⏱⏱–⏱⏱⏱⏱

PROJECT	COST	DIFFICULTY	DURATION
Easy Electronic Music	$-$$$	⚙–⚙⚙⚙	⏱–⏱⏱⏱⏱
Geeky Art Prints	$-$$	⚙	⏱–⏱⏱
Photoshop Your Kids into Their LEGO Kits (by Ken Jennings)	$	⚙–⚙⚙⚙	⏱⏱–⏱⏱⏱
Death Star Temari Ball	$$–$$$	⚙⚙⚙–⚙⚙⚙⚙	⏱⏱⏱–⏱⏱⏱⏱

"… are truly fun, inspired, and even educational … projects you can do with your kids." —*Wired*

978-1-592-40552-7

The instant *New York Times* bestseller

On Sale Now • Gotham Books

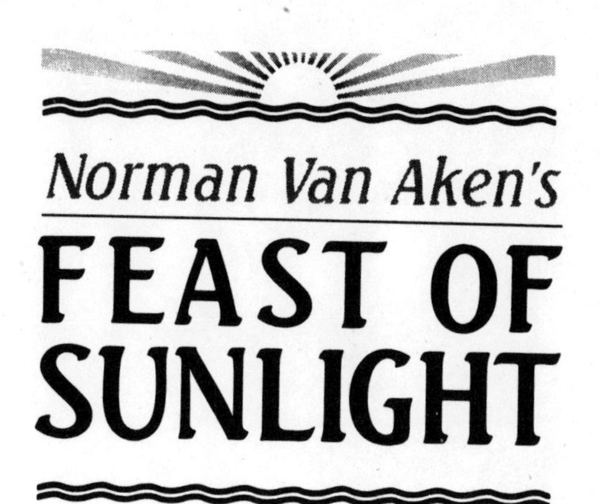
Norman Van Aken's
FEAST OF
SUNLIGHT